CAMBRIDGE MONOGRAPHS ON
MECHANICS AND APPLIED MATHEMATICS

General Editors

G. K. BATCHELOR, F.R.S.
Department of Applied Mathematics, University of Cambridge

C. WUNSCH
*Department of Earth, Atmospheric, and Planetary Sciences,
Massachusetts Institute of Technology*

J. RICE
Division of Applied Sciences, Harvard University

PRINCIPLES OF EARTHQUAKE
SOURCE MECHANICS

Principles of earthquake source mechanics

B. V. KOSTROV

Institute of Physics of the Earth,
Academy of Sciences, Moscow

SHAMITA DAS

Lamont-Doherty Geological Observatory
of Columbia University

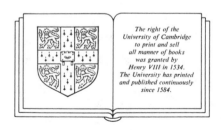

The right of the
University of Cambridge
to print and sell
all manner of books
was granted by
Henry VIII in 1534.
The University has printed
and published continuously
since 1584.

CAMBRIDGE UNIVERSITY PRESS

Cambridge

New York   New Rochelle   Melbourne   Sydney

Published by the Press Syndicate of the University of Cambridge
The Pitt Building, Trumpington Street, Cambridge CB2 1RP
32 East 57th Street, New York, NY 10022, USA
10 Stamford Road, Oakleigh, Melbourne 3166, Australia

Parts I and II of this work are a translation, with extensive
revisions, of a work by Dr. B. V. Kostrov first published in
the USSR in 1975 under the title *The Mechanics of the
Focus of Tectonic Earthquakes* (in Russian).

© Nauka Publisher 1975

Part III is original to this work.

English-language translation, revisions, and Part III
© Cambridge University Press 1988

Printed in the United States of America

Library of Congress Cataloging-in-Publication Data

Kostrov, B. V.

Principles of earthquake source mechanics / by B. V. Kostrov and
Shamita Das.

p.   cm. – (Cambridge monographs on mechanics and applied
mathematics)

Based in part on *The Mechanics of the Focus of Tectonic
Earthquakes.*

Bibliography: p.

Includes index.

ISBN 0-521-30345-1

1. Seismology.   I. Das, Shamita.   II. Title.   III. Series.
QE534.2K68   1988
551.2′2 – dc19   88-4249
                                                      CIP

British Library Cataloguing in Publication Data

Kostrov, B. V.

Principles of earthquake source mechanics.

1. Earthquakes. Mechanics
I. Title   II. Das, Shamita
551.2′2

ISBN 0 521 30345 1

Geology

CONTENTS

SYMBOLS

Symbols used frequently in the book are listed below; the list is not an exhaustive one. Occasionally the same symbol represents different quantities in different contexts, but we believe that no confusion should arise because the meaning of the symbol will be clear from the context. The equation number given with a definition identifies either the equation in which the symbol is defined or the equation in which it is first used.

a_0	Critical slip displacement in crack tip process zone
$\dot{a}(\mathbf{x}, t)$	Slip velocity vector at $(\mathbf{x}, t)$ on the fault (crack)
$\bar{a}_i$	Average or smoothed slip over fault
$\bar{a}_{i(\kappa)}$	Average of the ith component of slip over the fault due to the κth earthquake, $\kappa = 1, 2, \ldots, N$, $i = 1, 2$ [equation (4.5.1)]
$a_{i(\kappa)}(\mathbf{x})$	ith component of final slip (after earthquake) at $\mathbf{x}$ for the κth earthquake, $\kappa = 1, 2, \ldots, N$, $i = 1, 2$ [equation (4.5.1)]
$a_i(\mathbf{x}, t)$	Displacement jump across fault at $(\mathbf{x}, t)$, $i = 1, 2$
$a_{(n)}$	Normal component of displacement jump across fault (opening displacement)
$A^{\mathrm{P}}(t, \mathbf{m})$	P-Wave pulse shape at the observation point in the direction $\mathbf{m}$ from the source [equations (3.3.20) and (3.3.21)]
$A^{\mathrm{S}}_k(t, \mathbf{m})$	kth component of the S-wave pulse shape at the observation point in the direction $\mathbf{m}$ from the source [equation (3.3.21)]

$A^{SV}(t, \mathbf{m})$, $A^{SH}(t, \mathbf{m})$	SV- and SH-wave pulse shapes at the observation point in the direction $\mathbf{m}$ from the source [equations (3.3.22) and (3.3.23)]
$A_0^P(\mathbf{m})$	P-Wave radiation pattern for a point source [equation (3.4.5)]
$A_{0k}^S(\mathbf{m})$	kth component of the S-wave radiation pattern for a point source [equation (3.4.5)]
$A_0^{Q,\,n}$	Observed amplitude at station n for wave type Q at low frequency [equations (4.2.4) and (4.2.5)]
$A_{lm}^{Q,\,n}(\omega)$	Displacement spectrum for various components for wave type Q for station n
$A_{0lm}^{Q,\,n}$	Value to which $A_{lm}^{Q,\,n}(\omega)$ tends as $\omega \to 0$ [equations (4.2.4) and (4.2.5)]
b_i	Unit vector
$b_{i(\kappa)}$	Unit vector corresponding to the ith slip component for the κth source, $\kappa = 1, 2, \ldots, N$, $i = 1, 2$ [equation (4.5.1)]
c	Dimensionless constant, depending on fault geometry and material properties [equation (4.3.3)]
c_{ijkl}	Stiffness tensor [equation (1.3.17)]
d	Maximum dimension of fracture
$D_{ki}^{P,\,S}(\mathbf{m})$	P- and S-wave radiation patterns for the single-asperity model [equation (4.6.4)]
e_{klm}	Unit antisymmetric tensor [equation (3.5.11)]
E_q	Seismic energy on an earthquake [equation (4.3.7)]
f	Some function of one or more variables
$\tilde{f}$	Function defined by equation (3.7.1)
f_{MN}^a	Function defined by equation (3.7.6) and the equation preceding it
F_0	Positive constant independent of grid size representing the discretized value of $G_{\alpha\beta}(\mathbf{x}, t)$ at $(\mathbf{x}, t)$ [equation (5.2.7) and Appendix 2]
$F_P(t, \mathbf{m})$, $F_S(t, \mathbf{m})$	Functions of time whose dependence on the ray direction $\mathbf{m}$ determines the ratio of the actual source radiation to that of the concentrated dipole for P and S waves, respectively [equations (3.4.6) and (3.4.7)]
$F_{\alpha\beta}$	Discretized form of the kernel $G_{\alpha\beta}(\mathbf{x}, t)$ [equation (5.2.4)]
$\mathscr{F}^P(\mathbf{m})$	Function defined by equation (6.1.4)

G	Crack-driving force
G_0	Positive constant used in stabilizing the solution of the boundary–integral equation [(5.2.9) and Appendix 2]
G_i	Crack-driving force for Mode I ($i = 1$), Mode II ($i = 2$), and Mode III ($i = 3$) cracks
G_c	Critical G for fracture initiation
G_d	Critical G for fracture arrest
$G_{ik}(\mathbf{x}, t)$	kth component of the Green function (impulse response) of an elastic medium due to a point force directed along x_i, $i, k = 1, 2, 3$ (expressions for particular cases in Appendix 1)
h	P-Indicator function [equation (3.6.5)]
h'_{ϕ_0}	Derivative of h with respect to ϕ_0 [equation (3.6.13)]
$H(\ )$	Heaviside function
k	Coefficient of friction or bulk modulus
k_i	Stress intensity factors, $i = 1$ for Mode I (tension) crack, $i = 2$ for Mode II (inplane shear) crack, $i = 3$ for Mode III (antiplane shear) crack [equation (2.1.2)]
$k_{\mathrm{P}}, k_{\mathrm{S}}$	Wave number for P and S waves, respectively
k_{ic}	Critical value of the stress intensity factor k_i at which fracture occurs, $i = 1, 2, 3$
$k(\phi)$	Support function [equation (3.6.5)]
K_{ik}	Kernel of integral equation, defined by equation (3.2.19)
$K_{\alpha 3 k}$	Kernel of integral equation, defined by equation (5.1.2), $\alpha = 1, 2$
$K(\mathbf{x}, t), K(t)$	Smoothing kernels for slip distribution on fault
l	Some dimension of the fracture area
L_0	Length of initial defect [equation (2.4.11)]
L_c	Critical half-length of Griffith inplane shear crack
$\mathbf{m}$	(m_1, m_2, m_3), unit vector at source directed along the ray to receiver
$\mathbf{m}_1$	(m_{11}, m_{12}, m_{13}), unit vector at receiver directed along ray to the source
$\mathbf{m}'$	$(-m_2, m_1, m_3)$
$m_i^{\parallel}$	Component of $\mathbf{m}$ lying in the fault plane [equation (3.4.8)]
$m_i^{\perp}$	Component of $\mathbf{m}$ orthogonal to the fault plane [equation (3.4.8)]
$\mathbf{m}^{\parallel}$	Unit vector defined by equation (3.5.4)

$\mathbf{m}_P, \mathbf{m}_S$	$\mathbf{m}$ for P and S waves, respectively
m_{lm}	Components of the seismic moment tensor for a general source where the force is directed along the x_l axis and the arm of the couple along the x_m axis, $l, m = 1, 2, 3$, [equation (3.2.13); for an isotropic medium, equation (3.2.17)]
M_0	Seismic moment [equation (4.1.20)]
M_k	Mechanical moment [equation (3.2.15)]
$M_{lm}(\omega)$	Components of the seismic moment tensor for a concentrated dipole at frequency ω
M_{0lm}	Seismic moment tensor components as $\omega \to 0$ [equation (4.1.13)]
$M_{0(\kappa)}$	Seismic moment of the κth earthquake, $\kappa = 1, 2, \ldots, N$ [equation (4.5.3)]
$M_{0lm(\kappa)}$	M_{0lm} for the κth earthquake, $\kappa = 1, 2, \ldots, N$ [equation (4.5.14)]
$\mathbf{n}$	Unit vector, generally directed along the normal to the fault plane
OV	Dynamic overshoot in slip at the center of crack
p_i	Ray parameter; $= m_i / v_P$ and $= m_i / v_S$ for P and S waves, respectively
P_α	Geometrical constant related to an elliptical asperity
P_{MN}	Polynomial of degree $(M + N)$ [equation (3.7.6)]
$\mathscr{P}^P(\mathbf{m})$	P-Radiation pattern due to a concentrated dipole
q	Specific heat production [equation (2.2.3)]
q_i	Heat flux vector [equation (2.2.3)]
Δq	Heat flux from unit area of fault [equation (2.2.15)]
ΔQ_r	Radiation loss [equation (4.4.27)]
r	Final crack radius, or distance from crack tip [equation (2.1.2)], or distance from source to receiver, or radius of centered asperity in model of fracture of single-asperity on circular fault [equation (6.3.3)]
r_0	Initial crack radius
R	Distance from fracture initiation point to receiver or radius of outer circular crack in model of single-asperity failure [equation (6.3.1)]
R_P, R_{SV}, R_{SH}	Geometrical spreading factor for P, SV, and SH waves, respectively [equations (3.3.20), (3.3.22), and (3.3.23)]
$R(\omega, \xi)$	Form of the Rayleigh function in (ω, ξ) space [equation (3.5.13)]

S	Measure of fault strength $= (\sigma^u - \sigma^0)/\Delta\sigma$, or area of fault (crack), or boundary of elastic volume V
S_0	Surface area of source or value of tip element $S_{\alpha\beta}(x_1, x_2, t)$ at $x_1 = 0$, $x_2 = 0$, $t = 0$ [equation (5.2.13)]
$S_{(k)}$	Union of all grids with nonzero slip (discrete slipping area) [equation (5.2.13)]
$S_{(k)}^P$	Union of all grids influenced by disturbances at time $k\,\Delta t$ [equation (5.2.9)]
$S_{\alpha\beta}$	Numerical counterpart of $T_{\alpha\beta}$, $\alpha, \beta = 1, 2$ [equations (5.2.11) and (5.2.12)]
t, t_1	Times (e.g., times at source and receiver)
t_m	Source duration or maximum source duration
Δt	Temporal grid size in numerical solution of boundary–integral equation (5.2.1)
T	$v_P t/\Delta x$ or temperature [equation (2.2.17)]
T_P, T_S	P and S travel times [equation (3.3.13)]
$T_{\alpha\beta}(\mathbf{x}, t)$	Integral equation kernel defined by equation (5.1.4)
u_0	Applied shift at $\pm\infty$ in problems of failure of asperity on an infinite fault
$u_i(\mathbf{x}, t)$	Displacement components at $(\mathbf{x}, t)$, $i = 1, 2, 3$
$u^{\pm}$	Displacement components on $+$ and $-$ sides of fault [equation (1.3.25)]
$\dot{u}_i(\mathbf{x}, t)$	Velocity components
$\ddot{u}_i(\mathbf{x}, t)$	Acceleration components
$u_k^P(\mathbf{x}, t),$ $u_k^S(\mathbf{x}, t)$	kth component of P and S far-field displacements, $k = 1, 2, 3$ [equations (3.3.4) and (3.3.5)]
$u_k^P(\ddot{\mathbf{x}}, t),$ $u_k^S(\ddot{\mathbf{x}}, t)$	kth component of P and S accelerations, $k = 1, 2, 3$
$u_k^{SV}(\mathbf{x}, t),$ $u_k^{SH}(\mathbf{x}, t)$	kth component of SV and SH far-field displacements, $k = 1, 2, 3$
u_{SV1}, u_{SH1}	SV and SH displacements at the observation point [equations (3.3.22) and (3.3.23)]
u_{SV}, u_{SH}	SV and SH displacements at the source point [equations (3.3.25) and (3.3.26)]
U_{ik}	Green's tensor for elastic problems [solution of equation (3.1.6)]; for a homogeneous, infinite medium, equation (3.3.1)]
U_{ik}^P, U_{ik}^S	P- and S-wave contributions to U_{ik}
v	Fracture speed

v_1, v_2	Fracture speeds in x_1 and x_2 directions, respectively [equation (3.4.24)]
v_H	Speed of healing wave on crack
v_P	Compressional wave speed of elastic medium
v_R	Rayleigh wave speed of elastic half-space
v_S	Shear wave speed of elastic medium
v_{max}	Maximum fracture speed
V	Volume
$V^\pm$	Volume of integration on $+$ and $-$ sides of the fault
ΔV	Change in volume
w_0	Initial potential energy density
w^1	Final elastic energy density
w_e	Strain energy density [equation (1.3.22)]
$\dot{W}$	Rate of change of elastic energy within some volume [equation (4.4.13)]
$\mathbf{x}$	x_i, a Cartesian coordinate system, $i = 1, 2, 3$, or source coordinates
$\mathbf{x}_1$	Receiver coordinates
Δx	Spatial grid size in numerical solution of boundary–integral equation (5.2.1)
z	Unit vector at receiver directed along vertical
z_1	Unit vector at source directed along vertical
α, β	Subscripts with a value of 1 or 2
γ	Specific surface energy [equation (2.1.1)]
γ_{eff}	Effective γ
δ	Crack opening displacement at which cohesive forces at crack tip vanish [equation (2.1.7)]
δ_{ij}	Kronecker delta ($= 1$ for $i = j$; $= 0$ for $i \neq j$)
$\delta(\mathbf{x}), \delta(t)$	Dirac deltas
$\Delta(v)$	Rayleigh function [equation (2.3.22)]
ε	An arbitrarily small number
$\varepsilon_{kl}, \varepsilon'_{kl}$	Strain components, $k, l = 1, 2, 3$
$\dot{\varepsilon}_{(\kappa)kl}$	Average seismic strain rate due to the κth earthquake, $\kappa = 1, 2, \ldots, N$
$\overline{\Delta\varepsilon_{kl}}$	Average seismic strain rate tensor [equations (4.5.12) and (4.5.13)]
θ	Angle between direction to source and normal to fault [equation (6.2.1)], or $(90° - \theta)$ [equation (3.4.23)], or angle that applied shift at infinity makes with x_1 axis in elliptical-asperity failure problem

λ	Lamé constant
$\lambda_{\min}$	Minimum wavelength recorded [equation (3.3.40)]
$\lambda_{\mathrm{P}}, \lambda_{\mathrm{S}}$	P and S wavelengths [equation (4.1.3)]
μ	Modulus of rigidity
ν	Poisson's ratio
ξ	ξ_i is a Cartesian coordinate system, $i = 1, 2, 3,$ or $\xi_1 = \omega p_1$ and $\xi_2 = \omega p_2$ [equation (3.8.6)]
ρ	Density
σ^0	Initial stress (before earthquake)
σ^1	Final stress (after earthquake)
σ^u	Critical stress required for failure of nonideally brittle body
$\sigma^{\inf}$	Stress applied at infinity
σ^{stat}	Static frictional stress
σ^{kin}	Kinetic frictional stress
σ_B	Strength of laboratory rock sample [equation (2.4.15)]
σ_{r}	Average radiational friction
$\Delta\sigma$	Stress drop on crack (fault)
$\sigma_{ik}, \sigma'_{ik}$	Stress components
σ^0_{ik}	Component of initial stress
Σ_{ijk}	Stress tensor corresponding to Green's tensor U_{ik} [equation (3.1.9)]
$\Sigma(t)$	Fault area at time t
τ_0, τ_1	Pulse durations
τ_{ik}	Components of stress perturbations [equation (1.3.19)]
$\tau^{\pm}_{ik}$	τ_{ik} on $+$ and $-$ sides of fault [equation (1.3.25)]
ϕ	Polar angle [equation (3.6.5)]
ϕ_0	Polar angle [equation (3.6.12)]
Ψ^a_{MN}	Function defined by equation (3.7.7)
ω	Angular frequency
$'$	Derivative with respect to following variables
$\cdot$	Derivative with respect to time
$[\;]$	Jump of variable contained within the square brackets across any discontinuity

PREFACE

The practical goal of earthquake seismology is to prevent or reduce losses due to earthquakes by estimating the earthquake hazard at a given site or by forecasting the occurrence of the next strong event. The prevailing approach to these problems is to extrapolate data from the record of past events and apply that information to the future. Acceleration spectra of strong earthquakes are used to design structures, and unusual phenomena observed before earthquakes are considered possible indicators of future ones. This purely phenomenological approach is unreliable, due mainly to the lack of representative data. Recurrence times of strong earthquakes are fortunately very long, namely, several tens or hundreds of years, whereas the subject of seismology is only approaching its centennial. Obviously, an understanding of the earthquake generation process can partly fill the gaps in the available data, thereby making the practical conclusions more reliable. This is the purpose of earthquake source studies.

As is usual in geophysics, one can neither experiment with nor directly observe the earthquake focal region; one can only investigate its manifestations at the earth's surface. To convert these manifestations into information on the earthquake source process, one must have some general theoretical model that can then be fit to observational data. The history of the development of the source model is detailed in the Introduction. Making use of recent achievements in rock mechanics as well as progress in geophysical observations, the model has developed during the past several decades from the double-couple point source to the very sophisticated picture of the fracturing process at inhomogeneous faults. The complexity of the model reflects the complexity of the earthquake source process itself. It must be noted that oversimplification of the model may have unfortunate consequences for its practical application, as when catastrophic events are falsely predicted, mainly on the basis of theoretical speculations. At the same time, our current understanding of the complexity of the earthquake source makes it difficult to draw definite conclusions in view of the rather scarce observational information. The rapid development of geophysical data acquisition systems promises to improve the situation in the not too distant future. In any case, further refinement of the model is largely speculative. So although earthquake source theory is far from being complete or even consistent, it seems appropriate to summarize the more mature aspects of the state of the art. This is what this book is intended to do. More precisely, we confine the subject to the most basic principles of the theory and consider in some detail only the subject of earthquake dynamics, that is, the process of rupturing and the resulting seismic radiation during an earthquake, leaving aside the theory of earthquake preparation and interaction, which is at a less developed stage. We also include some mathematical tools for earthquake modeling. Our approach is purely deterministic, and no attention has been paid to the implications of the randomness of either the earth's structure or the source excitation process. Like Perry Mason, we consider only "possibilities," not "probabilities." The book deals only in concepts. Data-processing techniques used in seismology are mentioned purely for the purpose of illustration and are presented in only extremely simplified form.

The book consists of three parts, the first two of which are adapted from the Russian monograph *Mechanics of the Tectonic Earthquake Focus* by B. V. Kostrov, which was published in 1975 by Nauka, Moscow. These parts have been heavily rewritten and updated, and several new sections have been added. The third part is new. The book is meant primarily for use by graduate students and by teachers of earthquake source theory. We hope that the material will also aid the development of data-acquisition systems and data-processing techniques.

This book was written under Area IX of the U.S.–USSR Governmental Agreement in the Field of Environmental Protection and was partially supported by a grant from the United States Geological Survey during the very early phase of the work on the condition that the authors not accept royalties and the book be published by a nonprofit organization. We express our sincere thanks to all our friends and colleagues who encouraged us during the writing of the book, critically read early copies of the manuscript, and provided original copies of their figures. In particular, Kei Aki, Jim Rice, Jo Andrews, Jo Walsh, and Teng-fong Wong gave us their opinions, and Jim Dieterich and Barry Raleigh helped make it possible for us to write the book. We express our sincere gratitude to them.

INTRODUCTION

From a purely scientific point of view, an earthquake can be considered a source of information, the acquisition of which is the subject of seismology. The information conveyed by seismic waves consists of two fundamentally different parts: The first is created during the excitation of waves at the source of the earthquake; the second is produced by the conditions of wave propagation from the source to the station, that is, by the structure of the medium. Consequently, the interpretation of seismic observations requires the solution of two fundamental problems: the determination of the velocity structure of the medium and the determination of the earthquake source parameters. Although these two problems were recognized almost simultaneously, their roles in the history of seismology are different and they developed in different ways. Until recently, the investigation of earthquake sources received much less attention than the medium, and the relevant problems were, and to a large extent remain, little investigated.

There are several reasons for this situation. One is that structural seismology provides most of our information on the conditions prevailing at great depths and is the basis for many other branches of earth science. For a long time, the results from the study of earthquake sources seemed to be of more limited interest. In the past few decades, however, the determination of source parameters has become more important, possibly being a unique source of information about the stress conditions and present motions in the earth's interior.

Another reason is that the study of earthquake sources is much more difficult than that of the earth's structure. In fact, determining the velocity structure of the medium usually requires only kinematic parameters of seismic waves (travel times for body waves and dispersion curves for surface waves), which can be obtained with relative ease from seismograms and can be accumulated from many earthquakes for joint processing, being independent of the source process. In contrast, information on the motions and conditions at an earthquake source is contained mainly in the wave dynamics (waveforms) and is unique for every earthquake. Moreover, this information is partly (and sometimes totally) lost because of distortion by instrument and propagation effects. To eliminate these effects, the earth structure must be known beforehand.

Finally, the study of earthquake sources has encountered more serious difficulties in its theoretical aspects than has the study of the earth's structure. The inversion for the earth's structure is based on the theory of elasticity, the fundamental principles of which were formulated in the nineteenth century, and the corresponding mathematical methods were developed mainly in the nineteenth and early twentieth centuries. Of course, the refinement of numerical methods and especially the computer revolution have had a great impact on interpretation techniques, but this progress has taken place within the framework of classical concepts and has barely permitted broader application of classical results. Remember that the Herglotz–Wiechert formula was obtained in 1907 and Lamb's problem was solved in 1904. On the other hand, the attempt to achieve a theoretical understanding of earthquake source phenomena led to problems that in 1907 were impossible not only to solve but even to formulate precisely owing to the absence of adequate physical concepts for such phenomena. In fact, although the concept that earthquakes are the result of fracture of the earth's material due to tectonic stress, known as the elastic rebound theory, was formulated by Reid in 1910, the basis for analyzing this phenomenon – that is, fracture mechanics (sometimes called crack mechanics) – was not initiated by Griffith until 1921 and started developing vigorously only after the Second World War. Most of the dynamic problems in fracture mechanics proved to be unsolvable analytically, at least in closed form, by means of classical methods. Consequently, it was necessary to invent specific methods of solution for practically every problem. Nonetheless, fracture mechanics introduced certain physical concepts and methods of analysis that formed a framework in which to consider the phenomena of fracture nucleation, propagation and arrest in solid bodies and, particularly, at the earthquake source.

Naturally, destructive earthquakes forced seismologists to seek better insight into the earthquake process and to begin investigating earthquake sources with the limited resources available at the time, the lack of theory being compensated for by more or less explicit intuitive hypotheses as well as by arbitrary (and sometimes contradictory) simplifying assumptions.

Investigations of earthquake sources were shaped by the work of Reid, who formulated the theory of elastic rebound based on his study of the effects of the 1906 California earthquake. In the 1920s the regularity of the distribution of the signs of first arrival of seismic waves was discovered mainly by Japanese seismologists, and the concept of nodal planes was introduced. In his paper "Notes on the Nature of Forces Which Give Rise to Earthquake Motions," Nakano (1923) formulated the problem of finding the point source in the elastic medium for which the distribution of signs of first arrivals coincides with those observed for an earthquake and derived expressions for some dipole sources using the formulas or displacements due to a point force. This idea proved to be fruitful. It can be said that Nakano's paper initiated the quantitative study of earthquake sources. For a long time thereafter, effort was directed toward the development of this body-force equivalent.

It is clear that the determination of the body-force equivalent comprises, at best, only half of the problem, since it is also necessary to relate the characteristics of this equivalent to some physical concepts of the real earthquake source, namely, to the concepts involved in Reid's elastic rebound theory. This is reflected in the term "fault-plane solution," which denotes the body-force equivalent. Somewhat imprecise considerations regarding this problem led to a situation in which two contradictory point source models, namely, single couple and double couple, were proposed on the basis of the same physical (dislocation) model. This stage in the development of earthquake source investigations is well documented by the proceedings of two international symposia on earthquake source mechanisms. The first was held in Toronto in 1957 (The Mechanics of Faulting, with Special Reference to the Fault-Plane Work; see Hodgson, 1959) and the second in Helsinki in 1960 (A Symposium on Earthquake Mechanism; see Hodgson, 1960).

This dramatic situation was well described by Hodgson (1960) in his introduction to the proceedings of the second symposium: "At the time of the previous symposium the outstanding problem was that of selecting the appropriate mechanism and, since the theory had been established, this seemed to be simply a matter of further careful observations. Since that time much more theoretical work has been done, which has complicated rather than clarified the problem, and the observational results show little consistency" (p. 301).

Both of these point models explain the presence of nodal planes for longitudinal waves, one of which coincides with the fault plane at the source. However, whereas the radiation pattern due to a double couple is symmetric with respect to nodal planes, the single couple does not have such symmetry. This lack of symmetry would permit, if the single-couple model were correct, the determination of the actual fault plane from the signs of the S-wave arrivals. The impossibility of determining the actual fault plane in the framework of the double-couple model made an alternative interpretation of the axes of the point source orientation desirable. The conventional labels for these axes, P, T, and B, where P denotes pressure and T tension, were introduced to indicate that the axes coincide with the principal axes of the initial stress tensor causing the earthquake. Actually, these axes are the principal axes of the moment tensor of the equivalent point source. They characterize the distribution of fictitious body forces that would generate, in the medium without discontinuity, the same elastic field that is actually produced by the dislocation (fracture) at the source. Nevertheless, this interpretation is not entirely meaningless. In fact, assuming that the fault plane coincides with the plane of maximum shear stress, these axes would coincide with the principal axes of the stress tensor. This assumption, however, is not plausible from the physical point of view. For example, in the framework of the Coulomb–Mohr theory of strength, the fault plane would not coincide with the plane of maximum shear stress. Besides, when slip occurs along a preexisting fault, the plane of maximum shear stress (even if it initially coincided with the fault plane) may deviate from the fault plane on geologic time scales. Perhaps because of these considerations, this interpretation has not been widely adopted.

Discussion of the equivalent point source model concluded with the acceptance of the dislocation model. Later, by means of the dynamic Green functions, the double-couple model was related to the final slip distribution on the fault due to an earthquake. With these Green functions, it is possible to obtain an expression for the components of an elastic field at any point in terms of the displacement jump (the Burgers vector) distribution and history on the fault. The simplicity of this representation led to the development of a large variety of source models. The common feature of these models was that the distribution of the displacement jump on the fault surface was assumed arbitrarily (and in most cases as constant), and it was not clear to what extent the results depended on the choice of a particular distribution. This remark also applies to static dislocation models with a constant Burgers vector, which were proposed to describe the residual displacement at the earth's surface due to earthquakes. Essentially, such models represent a transfer to seismology of the theory of dislocations in an elastic medium – the so-called continuum theory of dislocations – developed to describe the behavior of dislocations in crystals. However, if, for dislocations in crystals, the constancy of the Burgers vector is due to the discreteness of the lattice structure, then for macrodislocation, which is what the earthquake source is, the assumption of the constancy of the Burgers vector, that is, a Volterra dislocation, is groundless and was accepted for the sake of simplicity. As Chinnery (1969) wrote:

> The theory... specifies in effect the displacement over a fault plane. As we have seen, it may be more useful physically to specify the stress acting across the face of a fault, and some improvements in the basic theory are needed to adequately include the effect of friction on the fault plane.
>
> However, in the present author's opinion, there is a definite limit to the amount of complexity that should be included in models designed for the study of faulting. The reason for this, as we have mentioned, is the general lack of adequate field data. The usefulness of any theory must be judged on its ability to predict new results, which may then be compared with observation. At the present time field data do not have the resolution to be able to distinguish between different theories which involve large numbers of variables. The complexity of the geology in the neighborhood of most active faults suggests that it may be a long time before there is a significant improvement in this situation.

Chinnery, here, is quite certain that the crack model for an earthquake fault has a stronger physical basis than the dislocational model with a constant Burgers vector. His only justification for the latter model is the insufficient accuracy of observations, which does not permit one to choose between the cracks-with-friction model and the Volterra dislocation model. It is true that geophysical observations, as such, are not likely to become sufficiently definite in the near future to permit a choice to be made between theoretical models. However, Chinnery's conclusion, that in that case any one of the observationally indistinguishable models is acceptable, does not follow.

To describe the fracture at an earthquake source as a crack, it is necessary to know the initial distribution of stress on the fracture surface before the earthquake and the laws governing the fracture propagation and interaction of the fault faces. Then, the distribution of the displacement jump on the fault becomes one of the unknowns. When this distribution is found, solving for other quantities reduces to the use of Green's formula, as in the above-mentioned dislocation models. Thus, the difference between modeling faults as cracks and as dislocations with a variable Burgers vector (Somigliana dislocation) is not as drastic as it may seem. When describing faults as fractures, one assumes some physical laws governing the fracturing and the external action applied to the fault (initial stress), and the motion along the fracture and within the surrounding medium is solved for, whereas in dislocational models, the motion along the fault is assumed. This is analogous to the two ways of describing the motion of a material point: kinematic when its trajectory is given, and dynamic when the forces acting on this point and the laws governing its motion are given but the trajectory is unknown. In the case of fracture, when it is described as a dislocation (displacement jump as a function of space and time is given, i.e., the trajectories of relative motion of all the initially adjacent particles), we will call it kinematic description, and when it is described as a crack, we will call it dynamic.[1] This terminology stresses the fact that both descriptions are related to the same thing, that is, fracture. At the same time, this discussion implies that the kinematics of fractures is an insufficient basis for the theory of earthquake sources because the displacement jump across the fault cannot be related to the physical laws governing the nucleation and propagation of fractures in a continuous medium and to the physical conditions that produce a particular fracture.

Since the 1960s, the dynamic description of the source has been generally accepted. With this description some new fundamental parameters of the source have been introduced into seismology. These parameters, such as stress drop, average slip, and fracture area, have replaced ones like source volume and strain release, which could not be formalized within the framework of the crack (fracture) model of the source. The notions that led to the introduction of parameters such as source volume were partly contained in the classic work of Reid (1910), as was the concept of the source as a fracture. A theory founded on these notions was developed by Benioff (1951). It is based on the assumption

---

[1] The term "dynamic" has two meanings: as the opposite either of "kinematic" or of "static." It is used here in the first sense, which leads to expressions such as "static problems for dynamically described fracture."

that fracture at an earthquake source occurs when the stress in a volume reaches the rock strength, the material breaks within the whole volume (the source volume), and the accumulated elastic energy in this volume is released. This energy was assumed to coincide with the seismic energy. Energy is released when elastic strain is transformed (totally or partially) into nonelastic strain as a result of the fracture. Since the elastic strain energy density is equal to the product of rigidity times the square of strain (for shear strain), the magnitude of the strain release, which is proportional to the square root of the seismic energy, can be evaluated. Benioff (1955) applied these considerations to the aftershock sequence of the 1952 Kern County, California, earthquake, which at the time was more extensively recorded than any previous aftershock sequence. This was the first instance in which portable and sensitive seismographs were set up on such a large scale within a few hours after the main shock in an earthquake area with good azimuthal distribution around the epicentral area. Benioff assumed that the source volume of the aftershocks coincided with the source volume of the main shock. Then the sum of the square roots of energies of aftershocks was proportional to the accumulated strain release, permitting the study of the dependence of strain release on time (Benioff plot). Identifying the strain release thus obtained with nonelastic strain within the source volume, Benioff considered the rheological properties of this volume and suggested an explanation for the very existence of aftershocks as well as for the time dependence of their magnitudes.

The mathematical simplicity and seemingly obvious physical basis of this theory make it appealing, so it is not surprising that it was readily accepted by practical seismologists. Benioff plots, not only for aftershock sequences, but for all earthquakes in seismically active zones and even the whole earth, were found to be a convenient means of analyzing time variations in seismicity. But it soon became clear that this theory could not be accepted without reservation. Even if one agrees that the fracture at the source occurs throughout a volume, the elastic strain energy must drop in the medium surrounding this volume as well, and consequently not only that volume in which fractures occur but the entire volume in which elastic strain energy is released should be considered to be the source volume. An attempt to modify Benioff's theory was made by Bullen (1953), who identified the source volume with the volume in which the stress was near the strength limit before the earthquake. Later, Bullen (1963, Chap. 15) extended this volume, considering the source volume to be that region where the major part of the energy released during the earthquake was accumulated. Actually he assumed that all of the elastic strain energy released during an earthquake was confined to a finite volume surrounding the source and that this volume was not larger by more than one order than the fracture region at the source.

At the same time, Tsuboi was considering similar ideas, but with a difference. In Benioff's theory, the energy of an earthquake is determined by the earthquake volume, so that a larger earthquake corresponds to a higher stress and a smaller one to a lower stress. Tsuboi (1956) suggested that the stresses for larger and smaller earthquakes are the same but that the volume differs and hence that larger earthquakes correspond to larger source volumes than smaller ones.

Benioff, Bullen, and Tsuboi all proceeded on the assumption that an earthquake leads to total release of the elastic energy accumulated in the source volume – that is, to total stress release in the volume. However, since the total energy release during an earthquake is simply related to the fracture area and the stress drop at the source, there is no need for any notion of source volume, and the concepts of source volume and strain release must be replaced by those of fracture area and stress drop.

In the early 1960s it started to become clear that the investigation of earthquake sources needed a more solid basis, namely, the physics of fracture in solid bodies. By this time the mechanics of brittle fracture, which is concerned with the development of fractures in solids, was fairly advanced and seemed to be a proper foundation for earthquake source theory. However, a simple transfer of the results obtained in fracture mechanics to seismology was impossible. The mechanics of brittle fracture was developed for engineering applications. In engineering, one is interested not in the process of fracture but in its prevention. Accordingly, in fracture mechanics attention was paid mainly to the theory of the critical (equilibrium) state of solids containing cracks. As for the dynamics of fracture – that of the fracture propagation process – very few experimental data were available and only the simplest problems had been solved. Moreover, under usual service conditions, the material is fractured, in most cases, in tensile mode, and fracture mechanics was concerned mainly with tensile cracks. For seismological applications, it was necessary to develop the theory of dynamic fracture propagation especially for shear fracture. Thus, there arose a need to generalize brittle fracture mechanics and to develop a method of solving dynamic problems of shear fracture propagation. The transfer of fracture mechanics to seismology required the reexplication of many mechanical terms.

In geophysics as a whole, forward problems are of minor interest, serving only to elucidate underlying physical phenomena. More important are inverse problems – problems that require the distribution of material parameters and motions in the earth's interior to be determined from surface observations. The inverse problem for the earthquake source has been formulated as one of reconstructing the displacement jump distribution and history over the fault surface at the source. The solvability of this problem was investigated, and two conclusions were reached: (1) Motion at the source is uniquely determined from its far-field seismic radiation (uniqueness theorem); (2) it is possible to construct a displacement jump distribution confined to an arbitrarily small area that produces seismic far-field radiation arbitrarily close to the observed one (instability theorem). These features, common to most inverse problems, imply that this problem cannot be solved without a priori information, in addition to seismic observations.

Chinnery's opinion, that the scarceness of observational data prevents one from distinguishing among sufficiently complex models of the source, has already been cited. Aki (1972a) writes: "Since the slip motion is a function of time and two space coordinates, a complete inversion is extremely difficult. The only practical inversion method is to describe the kinematics of rupture growth in a fault plane using a small number of parameters, and then determine those parameters from the seismograms." At first glance, these opinions are supported by the instability theorem. However, this theorem implies that, in principle, it is possible to construct two models with a finite number of parameters having arbitrarily different values, indistinguishable from one another with arbitrarily accurate seismic observations. Therefore, the additional constraints that have to be introduced for the practical solution of the inverse problem cannot be arbitrary, but they should follow from the physics of the source process. The latter is most adequately described within the framework of fracture mechanics.

Independent of this conceptual development, progress was made in practical seismology, which can be considered an approach to the solution of the inverse problem just described. The asymmetric Rayleigh wave radiation pattern from the 1952 Kern County, California, earthquake (Gutenberg, 1955) initiated the idea that the fracture speed was about the Rayleigh or shear wave speed of the medium. A similar fracture speed was found from surface waves of the great Chilean earthquake (Ben-Menahem and Toksöz, 1963). This led to the commonly accepted assumption that the size of the fracture area at the source is related to the pulse duration by the factor of Rayleigh or shear wave velocity. In 1966, Aki developed a method of determining the seismic moment and connected it with average slip and the area of fracture at the earthquake source. Then, from the seismic moment and fault size, the stress drop could be estimated. This development provided a unique possibility for estimating the stress conditions in the earth's interior. Nowadays, the determination of the seismic moment, stress drop, and so on has become routine, and together with fault-plane solutions comprises a firm support for global tectonic conclusions. In a sense, however, these developments went too far. Some nonphysical quantities, such as effective stress and seismic efficiency, were introduced and used broadly for drawing geophysical conclusions. The formalistic application of spectral techniques of seismic moment determination of small earthquakes became commonplace and produced a great amount of meaningless and misleading data. Sometimes, there was discussion of empirical relations between essentially identical quantities that differed only in notation and constant factors. (Some limitations of the applicability of these concepts and methods are discussed in Chapter 4 of this book.) Nevertheless, the impact of this progress on the development of earthquake seismology and its relation to other branches of geoscience cannot be overestimated.

In general, to obtain a complete description of the earthquake source it is necessary to determine the slip and stress fields on the propagating fault in space and time. Investigation of the properties of the inverse problem showed that this would be difficult, if not impossible. Instead, one can determine some overall features of the source – for example, the stress drop or slip averaged over the fault or the seismic moment tensor. If the principal axes of the source moment tensor do not change during an earthquake – that is, if the direction of faulting and the direction of slip on the fault do not change – then the time history of the moment tensor can be split into the (static) seismic moment tensor and a function describing its time dependence. This function is called the source time function. Since 1981 seismic moment tensors have been determined routinely and are now reported by Dziewonski and co-workers in the *Physics of the Earth and Planetary Interiors* as well as by the U.S. Geological Survey in its monthly listings in *Preliminary Determination of Epicenters*. The International Seismological Centre also plans to include these solutions in its monthly bulletins in the near future. Fault-plane solutions are now inferred not only from the signs of first arrivals, but also from the principal axes of the seismic moment tensor. Occasionally, differences are observed between these solutions. Such differences can sometimes be interpreted as having physical meaning; for example, the fault plane may have changed direction during propagation. Source time functions are now also a commonly determined quantity. The articles of

Ruff and Kanamori (1983), Helmberger (1983), and Ruff (1983) contain extensive bibliographies on the many earthquakes for which source time functions have been determined. The interpretation of source time functions in terms of physical processes occurring at an earthquake source requires intuition combined with mathematical modeling. Despite the importance of the role of the seismologist's intuition, such interpretations are of considerable importance and have provided strong support for models of heterogeneous faulting processes.

Because of the instability of the inverse problem discussed earlier, the study of forward problems for particular models of fracture at the source is of considerable importance. Essentially, solving these problems is the only way to obtain some insight into the detailed mechanics of the earthquake source as well as to explain some of the salient features of observations and laboratory experiments. In other words, the solution of forward problems provides a tool for understanding rather than for processing data. In most cases, forward problems cannot be solved analytically. This fact has given rise to the development of sophisticated numerical techniques and computer codes for dynamic crack propagation problems. An unusual situation also arose with respect to the relation between numerical and analytical solutions. Because of the practical limitations of computer time and storage, numerical solutions occasionally violate qualitative restrictions on the solutions of continuous analytic problems, such as the limitation on the speed of crack propagation, a point that we shall consider in some detail in Chapter 5. In such cases, however, the numerical solutions happen to be in better agreement with laboratory experimental data than do the analytic ones, although there is some doubt as to the reliability of the laboratory results. The reason is that in the analytic formulation the internal microstructure of material is totally ignored, whereas a numerical discrete formulation has microstructure, though artificial, and consequently the behavior of the numerical solution may correspond more closely to reality.

The intensive progress made in the study of earthquake sources, especially the development of methods for determining the physical parameters of fractures, necessitated the development, not of many individual models, but a single system of physical concepts, interconnected by such causal and logical relations, that would provide a framework for all actual data on the earthquake source – that is, the development of a physical earthquake source theory. As the basis for such a theory, continuum mechanics and fracture mechanics as well as physical notions on the source could be used. All these components were already present in an advanced form in the 1960s (at least for the dynamics of the source).

In the construction of mathematical models of earthquake sources in the 1960s, quantities like stress drop and fracture speed were assumed to be uniform over the fault. Of course, seismologists did not actually believe that these quantities were constant in the earth, and from the earliest days, when Haskell (1964) used a line source propagating at a constant speed to model the seismic radiation from earthquakes, it was clear that these assumptions were made only for the purpose of enabling the solution of problems. These simple models adequately explained some aspects of the data but could not fully explain others. For example, the very complex P-wave form of the 1964 Alaska earthquake (Wyss and Brune, 1967) cannot be explained by a simple Haskell model. With the development of sophisticated numerical methods and the advent of rapid computational facilities, seismologists began to consider more complex models and succeeded in further explaining observed features of seismograms. Many simplifications of course still remain. Current models assume that the fault is planar, though geologic observations often show large deviations from planarity. At present only the planar problems have been solved. As the quality of the data improves and more and more details of the observations fail to be explained by existing models, seismologists will be forced to develop methods to account for, say, the nonplanarity of the fault.

Nowadays complex distributions of the stress drop and strength on the fault are assumed in numerical models of the earthquake source, and the fracture speed is not prescribed but is determined as part of the solution using some fracture criterion. This complexity can be increased as desired and results in very chaotic fracture propagation patterns on the fault (Mikumo and Miyatake, 1979). This may, in fact, be a very realistic picture of faulting and leads to the fractal description of fault slip considered by Andrews (1980, 1982), based on Mandelbrot's (1977) theory of fractal surfaces. In Andrews's kinematic approach, the earthquake is a cluster in fractal slip history. The beauty of this theory is that both seismicity and a particular earthquake can be considered in the same framework. The fractal approach and the continuum approach are complementary methods for describing the behavior of earthquake faulting processes. But a fractal distribution of slip on a fault is not consistent with a smooth representation of the medium through which the waves propagate, and for a complete theory, the medium itself must also be described fractally. Clearly, to examine adequately the high-frequency radiation in the near field of a fault, a fractal or stochastic approach is essential. The ideas of Andrews and others have opened the door to a new and intriguing idea to be explored in the future. In this book, however, we shall pause on this side of the door and consider only a few of the simpler cases of nonuniform faulting and the resulting far-field radiation.

In order to apply fracture mechanical concepts to the earthquake source in a less speculative way, it was necessary to develop experimental techniques for determining the fracture toughness of rocks and to investigate experimentally possible forms of the constitutive behavior of frictional sliding of rocks. Some progress has been made in this direction during the past two decades and has led to a better understanding of some earthquake source processes (Byerlee, 1967; Dieterich, 1979; Rice, 1980; Wong, 1982; and others). A very recent "snapshot" of the state of the field in earthquake source mechanics can be obtained from the papers of the proceedings of the Fifth Maurice Ewing Symposium on Earthquake Source Mechanics held in 1985 (Das, Boatwright, and Scholz, 1986).

The present book is an attempt to summarize the fundamental principles and concepts of earthquake source theory. The subject is limited to the mechanical aspect of the theory and, particularly, to that of the earthquake in progress (coseismic stage of the earthquake). The earthquake preparation (preseismic stage) theory is extremely important in view of the earthquake prediction problem, and the postseismic stage should be included in a complete theory. The investigation of these stages, based on the study of time- and rate-dependent fracture phenomena, such as static fatigue and stress corrosion, is still at a very early stage. These are touched upon only slightly in Chapter 2.

The book consists of three relatively independent parts. The first presents the basics of continuum and fracture mechanics as applied to the earthquake source. The second is devoted to an analysis of the solvability of the inverse problem and related topics. The last part describes a numerical method for solving three-dimensional crack problems and determining the resulting seismic radiation, and presents some solutions that illustrate the general discussion of the previous parts as well as the power of the numerical approach. Some mathematical details of the numerical method are provided in two appendixes. In addition to the list of references at the end of the book, which contains only the material directly referred to in the text, we provide a list of suggested books and reviews that offer a complementary treatment of some of the topics discussed in the book.

Part I

CONTINUUM AND FRACTURE

1

BASIC MECHANICAL PRINCIPLES OF THE THEORY OF TECTONIC EARTHQUAKE SOURCES

1.1 Definition of the tectonic earthquake source; theory of elastic rebound

General remarks

The intuitive notion of the earthquake source is sufficiently clear, but at the same time the concept of it is quite vague. When defining the earthquake source for different applications, one emphasizes different aspects of the concept: the source as a point determined by first arrivals of seismic waves (the hypocenter), as the region where irreversible deformations occur during an earthquake, as the region where aftershock hypocenters are distributed, and so forth.

Let us begin our explication of the concept of the tectonic earthquake source with an analysis of the expression "earthquake source." It implies that the source is something different from the earthquake itself. By "earthquake" we mean the process of vibration of the earth's surface; in more general terms, it is the vibration of the earth or the propagation of waves. In this context the source is viewed as the source of seismic waves. It is something different from the material of the earth, or the medium, in which the waves propagate.

Next, we are interested in the source of a tectonic earthquake – that is, an earthquake that is produced by tectonic strain, when the energy radiated is due to the release of stress accumulated during tectonic deformation. Tectonic strain can be divided into elastic and nonelastic components. No potential energy is associated with nonelastic strain. Consequently, an earthquake occurs as a result of elastic strain drop. However, the total strain in the tectonically active region or in a part of it containing the earthquake source cannot decrease. Thus, the energy

17

released during an earthquake must be due to the transformation of elastic strain into nonelastic strain. This transformation may occur slowly, by creep or viscous or plastic flow, or rapidly, during an earthquake.

The fact that earthquakes occur suddenly implies local instability of tectonic deformation. In contrast to the instability of structures that is usually considered in structural mechanics, in which the shape of a structure becomes unstable while its material properties remain unchanged, the instability that we are considering here is related to the properties of the earth's material. To be unstable, the earth's material should be such that, under certain conditions, an increase in strain would lead to a decrease in stress (stress release). The widely accepted belief that stress release is associated with the formation of fractures (Reid, 1910) is based on observations of ruptures on the earth's surface that accompany earthquakes (faults, dislocations) and on the observation that earthquakes are confined largely to the vicinities of large geologic faults. In fact, there are no reasons to refute this opinion, at least for shallow earthquakes. Deep earthquakes occur at depths that are inaccessible to direct observation. Nonetheless, it can be shown that, even for these earthquakes, the process of stress release must produce fractures. We should note that, strictly speaking, only those earthquakes produced by the release of shear stress are to be considered tectonic ones. Actually, it is difficult to imagine that a change in volume plays a significant role in the process of tectonic strain. Of course, phase transitions with a change in specific volume are possible, and they may take place at great depths, but it is more appropriate to place earthquakes related to these phenomena in a special category. Let us call them "explosive" or "implosive" earthquakes to distinguish them from strictly tectonic ones. The quadrantal distribution of the signs of first arrivals, as observed in most cases even for deep earthquakes, implies that "explosive" earthquakes are rather rare.

Let us now consider the process of shear stress release in more detail. At first glance, it seems that the formation of fractures in the usual sense is not possible at the origin of a deep earthquake. In fact, consider a fault at a depth of a hundred to a few hundred kilometers. To produce slip on such a fault, it is necessary that the shear stress on the fault surface exceed friction. However, by multiplying the pressure at such depths by the usual coefficient of friction for rocks, one obtains an astronomical figure for the necessary shear stress, which is considerably higher than the yield stress of rocks. It is clear that slip cannot occur in such a way. On closer examination, however, one finds that this conclusion is based on the assumptions that at these depths Coulomb's friction law holds and that the coefficient of friction, at least its order of magnitude, is the same as determined in the laboratory for the rock samples. In turn, Coulomb's law, established only roughly for moderate temperatures and pressures, represents the friction phenomenon under the usual laboratory conditions. This is especially true of the constancy of the friction coefficient. Thermodynamic conditions (temperature and pressure) at great depths below the earth's surface differ markedly from those at the surface. Therefore, even if Coulomb's law is applicable to such depths, there is no basis for assuming that the friction coefficient under such conditions would be of the same order of magnitude as at the surface. Thus, the above reasoning does not imply that fault slip is impossible at great depths, but only that it is impossible according to the assumption of a constant friction coefficient and the validity of Coulomb's law. The next question is, What normal pressure should be multiplied by the coefficient of friction in Coulomb's law? Suppose for the moment that it can be assumed that the rocks at such depths are porous and the pores are filled by some fluid. This fluid may be juvenile water or a liquid phase of magmatic nature. In such cases it is obvious that the coefficient of friction must be multiplied not by the total pressure at this depth but by the effective pressure, that is, the difference between the overburden pressure and the pressure in the liquid phase.

This reasoning is based on the assumption of the presence of pore pressure at great depths, which is as baseless in this context as the assumption of the usual dry friction. Suppose now that frictional sliding is indeed impossible at such depths. It will be seen that even in this case the earthquake source is macroscopically equivalent to a fracture, under certain assumptions. These assumptions are that (1) tectonic earthquakes result from shear stress release, and (2) deep earthquakes actually occur. Accordingly, an earthquake is the stage of tectonic deformation at which the slow quasi-static process is replaced by the dynamic one. In other words, the earthquake is a loss of stability of the slow tectonic deformation, which is determined by the properties of the earth's material.

Consider a volume of material, say, in the form of a cube small enough to assume that strain ε and stress σ are homogeneous within it. For simplicity assume also that the strain and stress orientations do not change during deformation. Then the state of this volume can be graphically represented by the stress–strain curve. Let us assume that, at the beginning of deformation, stress also increases in the volume (or at

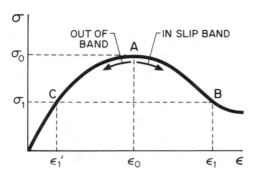

1.1 Unstable stress–strain relation.

1.2 Strain localization during instability.

least does not decrease). Then there will not be any earthquake, because the volume will always be in equilibrium with the neighboring parts of the medium that produce the strain (curve to the left of point A in Figure 1.1). If under some strain ε_0 the stress in the body starts decreasing under further strain, then instability sets in. In fact, consider the situation at point A in Figure 1.1. If the body experiences additional strain $\Delta\varepsilon$, the stress in it should fall as compared with the maximum stress σ_0. It will not counterbalance the action of the environment, and the strain will increase catastrophically unless the stress in the cube once again increases with strain ("strain hardening"). Without strain hardening, the strain will increase indefinitely. Thus, to produce a dynamic event, the stress–strain curve must have a descending part ("strain weakening"). Suppose now that, the entire cube being in a critical state at A, the additional strain $\Delta\varepsilon$ is confined to some narrow band (Figure 1.2). Point B (Figure 1.1) represents the state within this band. Because stress is continuous, it drops to a value of σ_1 all over the cube. Consequently, at this stage the strain outside the band decreases to a value ε_1 that is less than the critical value. Dynamic catastrophic increase in strain occurs only within the band. Physically, additional strain is caused by the inhomogeneity of the material. For example, the band might happen to have a smaller critical stress than the rest of the material, or the stress in this area might be greater due to inhomogeneity. Consequently, in an inhomogeneous material, the unstable deformation is necessarily confined to a band with thickness of the order of the scale of inhomogeneity rather than occurring in the bulk material.

If the earth's material is viewed as a continuous medium, small-scale inhomogeneities are neglected; that is, their size is considered to be

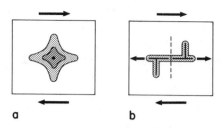

1.3   Development of region of unstable strain. (a) Beginning of the process;
(b) mechanism of locking of slip band.

physically infinitesimal. Therefore, the unstable strain responsible for an earthquake concentrates in an infinitely thin layer. This conclusion requires some refinement. In fact, it is more probable that the material will not reach the unstable state simultaneously along some layer, but first in a small volume (see Figure 1.3a). The shear stress in such cases would not decrease in the entire remaining volume; in some places, on the contrary, it would be concentrated. This would lead to the transition of other particles of the material into the unstable state. Simple analysis shows that, if the shear stress is applied as shown in Figure 1.3 by arrows, unstable strain should spread in the two directions of maximum shear stress concentration. However, if as a result of such sliding, the relative displacement becomes larger than the size of the initial inhomogeneity, then simultaneous movement along these two planes will not be possible and the slip will be arrested along one of the planes. As a result, a band of unstable strain (sliding) will occur along only one of these planes (Figure 1.3b). In the presence of strain hardening, further strain once again becomes stable until some new element attains the unstable state and a new band of sliding occurs. However, generally speaking, this would be another earthquake.

These considerations can be repeated mutatis mutandis for the case of rate-weakening instability with the same result, namely, that a tectonic earthquake is always associated with unstable strain of the earth's material, and such strain tends to localize in narrow zones (infinitely narrow when the material is described as a smooth continuum) that cannot be distinguished from cracks.

Hence, an earthquake source is a displacement discontinuity in the earth's material due to elastic (shear) stress accumulated during the process of tectonic deformation. In this definition, the distinction between the "source" and the "earthquake" is eliminated. The source is fixed as a specific thing, that is, a discontinuity in the earth's material.

Fractures (cracks) are stable or unstable depending on the distribution of external (initial) stress in space. A fracture is stable if its extension requires an increase in external load, being in equilibrium with the load. Unstable fractures spread at a fixed level of external load, and this propagation is fast (dynamic) because the equilibrium value of the load is a decreasing function of the fracture size.

Thus, a fracture is not always an earthquake source. It represents the source only during its dynamic propagation. Hence, the subject of earthquake source theory must be the study of the dynamic fracture process. Three phases can be distinguished here: the onset of dynamic propagation (initiation of fracture), the propagation of fracture itself, and fracture arrest. In this context, fracture propagation (the actual earthquake) appears only as a transition between its initiation and termination. These two are abstract instants of the earthquake process, but are themselves also transitions. The first extends from the initial quasi-static state of the medium to the process of catastrophic fracture, and the second extends from the fracture process to the final equilibrium of the medium. Under closer consideration the initial and final stages of dynamic fracture are seen not to be instants, but processes of finite duration due to time-dependent fracture phenomena (static fatigue, stress corrosion, etc.). These processes correspond to the preseismic and postseismic stages of an earthquake. Thus, tectonic earthquake source studies must consist of three basic parts: investigation of the earthquake preparation process, investigation of the earthquake in progress, and investigation of the postseismic relaxation – that is, the transformation of a propagating fracture as an earthquake source into a fracture per se.

The discussion in this section can be summarized as follows: The mechanics of a tectonic earthquake source is essentially fracture mechanics generalized to large spatial and temporal scales and to the conditions in the earth's interior. In particular, it is the mechanics of dynamic fracture propagation and seismic radiation.

Reid's theory of elastic rebound

The foregoing considerations clarifying the inner logic of the term "tectonic earthquake source" are somewhat abstract. It is necessary to make them more specific on the basis of data from a particular earthquake.

Reid's (1910) description of the 1906 California earthquake was the first and remains the most comprehensive detailed analysis of a specific earthquake. Never again in the history of seismology has the occurrence of a single earthquake advanced our understanding of the causes of earthquakes to such a great extent. Reid's conclusions, known as the theory of elastic rebound, deserve detailed analysis since all of the physical concepts of the tectonic earthquake source suggested since that time originate from his classic work. Not all of Reid's concepts are acceptable today, because in his day fracture mechanics did not even exist. Reid had to base his conclusions on the classical theory of the strength of materials.

It is worth quoting here several paragraphs from Reid's *The California Earthquake of April 18, 1906*, Vol. 2, which contain the original formulation of Reid's elastic rebound theory:

> The following is the conception of the events leading up to a tectonic earthquake and of the earth-movements which take place at the time of the rupture, as developed by the observations and study of the California earthquake and by the comparison of these observations with what has been observed in other great earthquakes.
>
> It is impossible for rock to rupture without first being subjected to elastic strains greater than it can endure; the only imaginable ways of rapidly setting up these strains are by an explosion or by the rapid withdrawal, or accumulation, of material below a portion of the crust. Both explosions and the rapid flow of molten rock are associated with volcanic eruptions and with a class of earthquakes not under present discussion; since earthquakes occur not associated with volcanic action, we conclude that the crust, in many parts of the earth, is being slowly displaced, and the difference between displacements in neighboring regions sets up elastic strains, which may become greater than the rock can endure; a rupture then takes place and the strained rock rebounds under its own elastic stresses, until the strain is largely or wholly relieved....
>
> The sudden displacements, which occur at the time of an earthquake, are confined to a zone within a few kilometers of the fault-plane, beyond which only the disturbances due to elastic vibrations are experienced. The distribution of the distortion of the rock at the time of the California earthquake shows that the elastic rebound and consequently the elastic shear was greatly concentrated near the fault-plane and was much reduced in intensity at even short distances from it; this concentration of the shear brought about a strain sufficient to cause rupture after a comparatively small relative displacement of the surrounding regions; if the shear had been more uniformly distributed over a wider region, a larger relative displacement would have been necessary to cause a rupture and there would have been a greater slip at the fault-plane. Therefore, altho [*sic*] it is quite conceivable that regions at a distance apart of, let us say, several times 20 km., might be relatively displaced and set up a state of elastic strain in the broad intervening area, it would be necessary that the relative displacements

of the distant regions should be at least several times 6 meters, in order that the strain should become great enough to cause a rupture; and if the strain were less concentrated than it was in California, the relative displacements would have to be greater still. It is only in the case of very large earthquakes that a slip as great as 6 meters occurs; and we may therefore infer that it is only in the case of large earthquakes that the sudden elastic rebound is appreciable as far as 8 or 10 km. from the fault-plane.

The rupture does not occur simultaneously at all parts of the fault-plane; but, on account of the elastic qualities of the rock, it begins in a very limited area and spreads at a rate not exceeding the velocity of compressional waves in the rock....

We know very little about the interior of the earth or of the origin of the forces which produce such great changes at the surface. Great thrust faults exist which indicate tangential compressions; and normal faults, which indicate expansion. Great uplifts have occurred unaccompanied by compressions, due, apparently, to vertical forces; and the California earthquake has emphasized the existence of horizontal drags below the crust. Future study may reveal forces applied in other ways; but it is not going too far to say that whenever ruptures occur, they result from elastic strain, and the sudden movements produced are merely elastic rebounds; and moreover, except in the case of earthquakes connected directly with volcanic action, the strains have not been set up suddenly, but are gradually developed by the slow displacements of adjacent areas. (Reid, 1910, Vol. 2, pp. 29–31)

In the first two paragraphs, Reid draws a distinction between strictly tectonic and other (volcanic) earthquakes, which we called "explosive" and "implosive" earlier. Furthermore, he states that earthquakes occur as a result of ruptures that are produced only when stress attains a value "greater than it [the rock] can endure." This value, usually called the strength, was assumed to be a material constant in Reid's time. It has been established in fracture mechanics that the strength is determined not only by the properties of a material but also by the size and distribution of defects in it. That fracture tests give approximately constant strength under the same confining pressure is due to the selection of samples without appreciable defects, so that only inherent defects influence the strength value. In the case of the earth's material, not only microcracks of the crystalline grain size of the material but also active faults hundreds or thousands of kilometers in length as well as multiple fractures of practically all intermediate sizes are present. Thus, the term "strength of the earth's material" becomes essentially meaningless and earthquakes may occur not only because of increasing stress but also because of growth in the number and size of defects, that is, the initial fractures.

Moreover, Reid associates the occurrence of fractures with the unloading of surrounding rocks, so that the elastic strain is "largely or wholly relieved." Further reading of the text shows that Reid was inclined to prefer "wholly relieved."

The next two paragraphs deal with the concentration of elastic strain (stress) near the fault plane. Here Reid comes to two conclusions. First, geodetic data show that the strain arising as a result of an earthquake is concentrated near the fault within a zone roughly 20 km wide; assuming that this strain differs from the accumulated initial elastic strain only in sign (total strain release), Reid concludes that this latter is concentrated in the same zone. Second, from laboratory values on the strength and shear modulus, Reid concludes that the strain, corresponding to the 6-m relative displacement of two points separated by a distance of 20 km, is just sufficient to achieve the strength value. If the zone where elastic strain was concentrated were wider, the slip would have to be proportionally greater.

Both of these conclusions are incorrect. Total stress (elastic strain) release is possible only for a fracture extending infinitely in all directions. Actually, faults have limited size in the horizontal and (especially for strong shallow earthquakes) in the vertical directions. The movement along the fault is thereby constrained by the adjoining unbroken parts of the medium, so that the magnitude of slip and size of the region in which appreciable residual displacements on the earth's surface take place are determined not by the size of the region under stress but essentially by the dimensions of the fault itself. In particular, for a given fault, this region must be determined mainly by the fault depth. In general, the distribution of strain due to a fracture is determined completely by the configuration of the fracture area, by the stress drop on this area, and by the properties of the medium, so that the distribution of strain does not depend on the spatial distribution of initial stress off the fault. In fact, the residual displacement distribution obtained for cracks in a half-space under homogeneous stress has been found to agree with the geodetic data for the 1906 California earthquake from which Reid inferred his conclusions (Knopoff, 1958; Berg, 1967; Walsh, 1968).

Reid's second conclusion is based essentially on the first one and on the laboratory values of rock strength. These values, as mentioned earlier, are not applicable to the real material of the earth.

In the next paragraph, Reid formulates a very important principle: that fractures do not occur instantaneously, but propagate at a velocity not exceeding the velocity of compressional waves in the medium. The basis for this conclusion (other than his observations) is the fact that, as already stated, the medium is "microinhomogeneous" and the critical state cannot be reached simultaneously at all points of a volume or area, even if the medium is "macroscopically" homogeneous. Instead, fractures that originate at one place cause a redistribution of stresses in the medium and stimulate fracturing in neighboring parts. Obviously, this redistribution cannot occur at a velocity higher than the velocity of compressional waves of the medium.

Although the impossibility of instantaneous formation of fractures in the entire volume was clear to Reid, such models were sometimes studied later. Assuming instantaneous fracture formation or at least supersonic fracture propagation greatly simplifies corresponding problems and allows the analytical treatment of more complex models. This, perhaps, motivated such studies. To justify the formulation of problems with supersonic fracture velocity, it was occasionally said that the foregoing arguments did not prove the impossibility, but only the low probability of simultaneous attainment of a critical state by different elements of the medium. Such a probability (the estimation of which is almost impossible) ought to be so small as to agree totally with an example given by Boltzmann in another context: This probability is comparable to the probability of fire breaking out simultaneously in most houses of a big city.

Finally, Reid discusses the possibility that the occurrence of several fractures might serve as the source of a single earthquake and through examples shows that, though rare, such a possibility does exist.

Thus, from Reid's theory, we retain the concept of the source as a fracture induced by tectonic shear stress and the arguments on the limitations of fracture velocity but reject the conclusions as to the concentration of initial elastic strain near the fault as well as the conclusions following from the application of the classical theory of strength of materials. Unfortunately, in presentations of Reid's theory, attention is often focused on these two unfounded aspects. In particular, Benioff's theory, which was discussed in the Introduction, is actually based on this part of Reid's theory.

Formal definition of tectonic earthquake source

The conclusions drawn from the foregoing analysis can be formulated in the form of the following five assumptions, which constitute the formal definition of a tectonic earthquake source. In fact, they are simply a condensed formulation of the theory of elastic rebound, without obsolete assumptions.

1. A tectonic earthquake source is the fracture of the earth's material along a (plane) surface.
2. Fracture results from (shear) stress, which accumulates during tectonic deformation, and leads to total or partial stress release over the fracture area.
3. Fracture is initiated over a small area and then propagates at a velocity not exceeding the velocity of longitudinal waves (causality principle).
4. Fracture corresponding to the tectonic earthquake source is a shear fracture; that is, the normal displacement jump is negligible.
5. The material surrounding the fracture surface remains linearly elastic.

The first assumption is, perhaps, too rigid. The fracture surface may not be plane, and the formation of several fractures during a single earthquake cannot be ruled out. It is unlikely, however, that all these fractures would appear simultaneously (see assumption 3, the causality principle), so that this hypothesis is justified at least for the initial stage of an earthquake. In what follows, wherever possible, this assumption is avoided.

The assumption as to the absence of tensile fracturing is also unnecessarily rigid. Generally speaking, the possibility of tensile fracture during earthquakes cannot be ruled out. For example, Reid noted the possibility of the formation of "gaping fissures" due to tensional stress or bending. However, such cases are possible only for the shallowest earthquakes and are expected to be rare. In the case of deep earthquakes the transition to an unstable phase of strain might be accompanied by a change in the specific volume of the material, due, for instance, to melting, and in the external description of such slip bands as fractures this change of specific volume is equivalent to opening. But in this case the tensile component of displacement must be small compared with the shear component. Consequently, for all tectonic earthquakes, this hypothesis is expected to hold.

The fifth assumption is to be understood in the sense that the response of the medium to rapid disturbances during an earthquake is linearly elastic. In fact, during slow tectonic strain the earth's material experiences plastic flow. Moreover, high stress and correspondingly large nonlinear and irreversible strains should appear near the fracture, especially near its edges. This assumption implies that the inelastic zone is sufficiently small to be considered physically infinitesimal and to be incorporated into the fracture surface.

The foregoing definition of the source, like any other formal definition, is a distortion of reality. But it is a precise definition – not a vague notion – that can be transformed into quantitative language. Furthermore, it is a minimal definition – that is, it includes only those assumptions that necessarily follow from the analysis of the concept itself and from observations.

In order to translate this definition into a quantitative formulation, it is necessary to clarify the terms involved: "continuum," "fracture," "stress," "strain," "elasticity," and so on.

1.2   Kinematics of a continuous medium

Few would disagree that the theory of elasticity constitutes the theoretical basis of seismology. Nevertheless, it is worth examining the assumptions and concepts of the theory of elasticity that are used in seismology. On the one hand, laboratory rock samples possess, to some extent, the properties of an elastic medium as defined in a mathematical model. On the other hand, observations of the propagation of seismic waves show that this propagation is described well by the theory of elasticity. One can say that rock samples are elastic. Similarly, one can say that seismic waves are also elastic or, more accurately, that they are waves in an elastic medium. Hence, with regard to seismic waves, both laboratory rock samples and the earth's material can be described by a mathematical model of elastic continuum. But do they represent the same elastic medium? Before answering this question, let us see what exactly is meant by a "continuous medium."

Concept of a continuous medium

Let us consider a body, say, a rock sample or some volume inside the earth. If we are interested only in the motion of this body as a whole, we are in the domain of classical mechanics, the mechanics of a rigid body. But in the course of its motion, the body might change its shape. A description of this change and its relation to external actions and properties of the body itself constitutes the subject of mechanics of a continuous medium. If the shape of the body changes during its motion or, more precisely, if we cannot neglect these changes for some reason, we cannot speak about the motion of this body as a whole but only about the motions of its parts or how these parts move relative to one another. We try to describe the motion of these parts in terms of classical mechanics, neglecting the fact that they are capable of changing shape. But actually the shape of each part of the body, together with that of the whole body, changes during motion. Consequently, we are left with no alternative other than to examine each part, which in turn consists of several parts. That is, even smaller parts of the body have to be considered in an attempt to describe their relative motion. It is quite clear that, whatever small parts of the body are chosen, this contradiction will remain and will force us to select smaller and smaller parts. If this process of selecting smaller and smaller parts of a body could be continued indefinitely so that it would be possible to select parts whose dimensions were smaller than any given value, we would have a continuous body or a continuous medium. The process itself would lead us to the concept of infinitesimal particles, or simply "particles of the medium" of which (according to our understanding) the body was composed. But for physical bodies, rock samples, for example, this process of dividing the body into parts in such a way that each of its parts is a solid body cannot be conducted indefinitely, if only because of the molecular structure of matter. Thus, a physical body can never be a continuum, and hence we cannot speak of its infinitesimal particles as they are understood in solid mechanics.

Consequently, in physics, continuity itself has a different meaning than it would in a treatise on continuum mechanics; the concept of a continuous medium is only a model for physical bodies, and infinitesimal particles are only a mathematical model for real parts of a body. These real "sufficiently small" bodies that are represented by infinitesimal particles are called "physically infinitesimal." A physically infinitesimal particle is that part of a body that plays the role of an infinitely small one. For this purpose, its dimensions should be insignificant in some particular respect. With regard to strain, the dimensions of the particle ought to be such that a change in its shape could be easily described for a particular problem, the essence of this change being conserved if the dimensions varied several times. For example, a volume containing only a few molecules cannot be considered physically infinitesimal, because by reducing its size a few times it is possible to obtain a volume that does not contain any molecules at all. If we take a volume containing many molecules, it is meaningless to talk about the motion of an individual molecule. We must talk instead about some average motion of this element and its configuration, for the discrete structure of matter does not prevent us from considering this volume to be physically infinitesimal. If, however, we are interested in knowing how the different parts of a body change shape, then this particle should not be so large as to obscure the strain variations in it. Imagine that, within measurement error, all parts of the body have similar properties and that every part changes shape similarly. Then by selecting smaller and smaller parts of the body, we would observe that their relative positions changed uniformly, and we could assume that however small were the parts of the body we considered, this would still be true; that is, we could imagine that the dimensions of its parts could be reduced indefinitely in spite of it being clear that the relative positions of individual molecules changed in different ways in different parts of the body and that in a sufficiently small volume it would be impossible to speak of the relative position of anything. What is important to us is that in a definite range of particle sizes their relative positions vary to the same extent, so that we can simply assume that this would also hold for indefinitely smaller particles. Though actually this is wrong, it is irrelevant for our purpose as long as we are not interested in the relative positions of too small particles, for example, since we cannot measure them.

Now, if sufficiently large parts of a body change their shapes in different ways, we can expect that by selecting smaller parts we will reach a situation wherein the relative position of particles within each such part will not essentially depend on the size of the particles unless they become too small. Then we can again assume that a further decrease in the size of these particles would be pointless.

Using mathematical terminology, it can be said that, if the concept of an infinitesimal element in the mathematical theory of a continuous medium is based on the process of approaching a limit in the strict sense, the concept of a physically infinitesimal element is based on the process of asymptotically approaching a limit. This is similar to the summation of an asymptotic series when a sufficiently large number of terms in the series is taken, so that their sum changes little if a few more terms are added. But we should not take too many terms, because in that case no sum can be obtained as the series diverges.

Next, we observe that, when dealing with seismic oscillations, it is meaningless to speak of relative positions of the parts of a medium in regions whose dimensions are much less than the wavelength. When we consider seismic waves with wavelengths of the order of one to hundreds of kilometers, the size of the particles that we can consider to be physically infinitesimal is of the order of hundreds of meters. At the same time, when dealing with laboratory samples, we must consider a few centimeters or millimeters as physically infinitesimal. When introducing a continuous medium as a model of a real body, we should always keep in mind that an infinitesimal particle is in fact the physically infinitesimal one, which should be not only sufficiently small, but also large enough to be representative; that is, the property of interest should not depend on its size within some limits. Consequently, to describe the behavior of material relative to the propagation of seismic waves we could not select elements that were only a few centimeters in size, because they would not be representative of volumes with a size of the order of hundreds of meters. In this respect, laboratory rock samples are modeled by an altogether different elastic medium than the material in which seismic waves propagate.

Continuous description of the earth's material

To make the meaning of this last conclusion clearer, we shall provide two simple examples. Suppose we have a spiral steel spring. We must consider as particles such parts of it as can still be springs – that is, sections containing at least a few turns. Thus, we arrive at the concept of a one-dimensional continuum where the positions of points depend on a single coordinate. At the same time, the spring is a piece of steel wire, and we could be interested in the strain of the wire itself. Here, as particles we must take such parts of the spring as are still parts of the wire but can be considered linear. In this case, however, to determine the position of the particle, in addition to the position of its center of mass, we must define its orientation in space. In this case, we again obtain the one-dimensional continuous medium, namely, the generalizations of it when the shape of particles and their orientation are taken into account, that is, the model of an oriented continuum. If we are interested in knowing the strain of the spring as a piece of steel, we must take particles that themselves are pieces of steel whose shapes are immaterial. Therefore, we have a three-dimensional continuous medium that is again classical (not oriented). Of course, in all three cases we are dealing with the same physical object, that is, the steel spring, but its strain is described by entirely different values, and its properties as a spring are quite different from the properties of a wire or a piece of steel, even though they are determined by the latter.

Consider, as the second example, a rock sample that is a polycrystalline body. By studying the strain of samples of different sizes and shapes, we would arrive at the conclusion that shape and size are immaterial in the sense that it is irrelevant to us whether we choose samples of 5 or 20 cm for experiments. Examining particles of such an order of magnitude, we would obtain the model of the rock as a homogeneous continuum. However, the rock consists of crystals of different kinds, as well as pores and flaws. Therefore, in order to describe the strain within crystals it is necessary to select particles whose size is smaller than the grain size of the rock. Consequently, the strain in the sample at one and the same point, described at these two levels, would be quite different. In this case, the strain on the grain level is referred to as "microstrain."

Let us now consider a larger volume (tens to hundreds of kilometers) within the earth, one consisting of blocks with faults, voids, and inhomogeneities on different scales. If we want to discuss the strain in this volume on a large scale, it is necessary to consider particles whose shape and size are immaterial, that is, that contain many structural elements such as blocks and faults. The measure of strain that we assign to the given point may be significantly different from the strain that we would have obtained had we selected particles a few meters or a few tens of centimeters in size and even more if the size of the particles were a few centimeters.

Here the concepts of "microstrain" and "macrostrain" are already insufficient, because we have structural elements of different orders, so that with respect to the tectonic strain of a large region, there will be many "microstrains" of various levels down to the level of polycrystalline grains. That these arguments are not merely pedantic but important for a correct understanding and organization of geophysical observations is seen from the consideration of slow movements. Tilt measured with a tiltmeter partially represents the strain of the earth's surface over an area of the size of the tiltmeter. It is clear that these measurements might be representative of an area with dimensions of at most a few meters (maybe several tens of meters). Obviously, from these measurements we cannot conclude anything about the tectonic strain over regions measuring a few kilometers or several tens of kilometers because such regions, in general, are represented by a different continuous model with different measures of strain. Geodetic measurements, which are representative of larger areas, are a better means of studying tectonic strains.

Detailed seismic studies (deep seismic sounding, seismic prospecting) have shown that the earth has a much more complex structure than indicated by the analysis of teleseismic observations. In particular, a large number of thin layers and interfaces have been found where seismological observations had suggested that the earth's material was a "smooth" medium. Hence, even here the model of the medium would depend on the scale and particularity of the observations. In other words, the earth's material is one medium for seismic prospecting, another for deep seismic sounding, and still another for earthquake seismology, in the same way that a spring is one medium for a designer of seismographs, another for a spring designer, and a third for a metallurgist.

Of course, the differences between these media (earth's material for researchers in petrology, rock mechanics, seismic prospecting, deep seismic sounding, earthquake seismology, and structural geology) are not absolute; only the details of the descriptions of the same real objects differ, the properties of the media being not so drastically different as the properties of a spring and a piece of steel. However, this difference in properties, as we will see later when considering fracture, and in the example below, is of major significance.

Accounting for this scale dependence of the continuous medium that is used to model the earth's material may perhaps explain the paradox concerning precursory changes of travel times before earthquakes. Namely, laboratory tests (Nur, 1972; Hadley, 1975, 1976) and observations of travel times of seismic waves from earthquakes (Nersesov, Semenov, and Simbireva, 1969; Semenov, 1969; Aggarwal, Sykes, and Armbruster, 1973; Whitcomb, Garmany, and Anderson, 1973) reveal appreciable changes in seismic velocities before fracture or sometimes before an impending greater earthquake. Laboratory experiments suggest velocity changes as large as 20 percent or more. Teleseismic travel-time residuals sometimes are observed to change up to 1 sec near the epicenter of an impending earthquake (Wyss, 1975). If these variations are due to changes in seismic velocities in the crust, these changes ought to be about 10 percent or more. Variations in travel-time observations from local earthquakes are less definite, but are well documented.

In contrast, very accurate measurements of travel times from quarry blasts did not show any precursory changes even though the seismic ray crossed the focal zone of a future earthquake of magnitude 5 (Boore, McEvilly, and Lindh, 1975). Another set of accurate travel-time measurements carried out on the Kamchatka coast from explosions in the sea several tens of kilometers distant showed negligible variations of the order of 8×10^{-2} sec (which is near the limit of accuracy of experimental measurements) even before an earthquake of magnitude 7.2 (Myachkin, 1978; Myachkin et al., 1985a). These experiments are considered very reliable because the origin time is known very accurately.

These observations have cast great doubt on the very existence of precursory velocity changes and have created a tendency to discard the corresponding seismological observations.

To examine this paradox (Myachkin et al., 1972, 1985b; Dolbilkina et al., 1979), let us take into account the fact that the frequencies involved in teleseismic travel-time determination are much lower than those for local earthquakes, which in turn are appreciably lower than those for blasts. Accordingly, the representative particle size and the corresponding continuous models of the media are different for these different waves. One might say that long waves from distant earthquakes propagate in a different medium than waves from local earthquakes or blasts. The usual explanation of the supposed precursory travel-time changes is based on the assumption that, before an earthquake, many cracks, active faults, and so on appear in its focal region. When describing material with such defects on a scale larger than their sizes and the distances between them, one obtains a continuous elastic medium with reduced moduli. In laboratory experiments, these precursory cracks are detectable, and the theoretical predictions agree fairly well with observed velocity changes. The model of the material obtained in this way can be applied to the description of wave propagation only for waves whose wavelength is much longer than the size of the representative particle – that is, much longer than the sizes of the defects and the distances between them. When the wavelengths are much shorter than this size, however, the material cannot be described by this smooth model, and when considering the propagation of such waves one must take into account all the individual defects. These short waves propagate in between the defects in the medium with unreduced moduli (the elastic matrix). Of course, the moduli of the latter are influenced by the presence of defects that are much smaller than the wavelength of the short waves under discussion. Hence, if a direct wave of such short wavelength arrives at the observer, the travel time will not be influenced by the appearance of the large defects. Instead, it will be scattered by these defects, and the arrival of these scattered waves will probably change the observed waveforms. Magistrale and Kanamori (1986) reported such a change in the waveform of Nevada nuclear test site seismograms recorded in the source region of the North Palm Springs, California, earthquake of July 8, 1986.

Consequently, if cracks or other defects appear in the vicinity of the hypocenter of an impending earthquake and their sizes are proportional to the magnitude of the earthquake, then sufficiently long seismic waves will be delayed without a change of waveform, whereas the waveform of short waves will be distorted without any variation in travel time. In fact, during the above-mentioned Kamchatka experiment, appreciable changes in waveforms were observed before the November 24, 1971 ($M_s = 7.2$), Kamchatka earthquake. So the explanation of this unresolved paradox should be sought in the difference in the wavelength of the sounding waves relative to the size of the preparatory defects in the source zone. Possibly, the failure to detect precursory variations is due to the fact that the measurement is too accurate! This explanation of the paradox, which is based on the wavelengths involved, is essentially the same as that in optics for x-rays, which are not refracted by a glass prism in contrast to visible light rays.

Quantitative theory of strain

The quantitative formalism presented in the remainder of the chapter is included for convenience (particularly of notation). It is far from being complete, and it is not as general as it could be. Most of it is quite standard material and can be found in textbooks on seismology (e.g., Aki and Richards, 1980) or on the theory of elasticity (Love, 1944). A more profound presentation of continuum mechanics can be found in Green and Zerna (1954), Nowacki (1963), and Truesdell (1965–6) to name just a few of the many textbooks available on this subject.

In order to represent the configuration of a body at an initial instant t_0, the positions of all physically infinitesimal particles of that body must be specified. Since the sizes of such particles are insignificant for us, their position can be specified by the coordinates of their centers of mass in some Cartesian system of coordinates x_1, x_2, x_3. Each such particle is supposed to be identifiable at any other instant t, even if it is at some other place and has changed its shape. Neglecting the size and shape of the particles, one may consider them to be "points" of the body defined only by their coordinates at t_0; that is, these coordinates are used as "names" of the points of the body. To distinguish them from spatial (current) coordinates, it is convenient to introduce the special notation ξ_i (Lagrangian coordinates) for them.

Consider the position of a particle having Lagrangian coordinates ξ_i ($i = 1, 2, 3$) at some instant t. This position depends on the initial position ξ_i and the time t, that is,

$$x_i = x_i(\xi, t), \qquad i = 1, 2, 3 \tag{1.2.1}$$

where the triplet (ξ_1, ξ_2, ξ_3) is represented by ξ.

Now, a description of the material as a continuous medium consists of the assumption that at every point in the volume, occupied by the body at every instant, there will be some infinitesimal particle and that the functions (1.2.1) constitute a continuous and one-to-one mapping of the volume occupied by the body at the instant t_0 on the volume occupied by it at the instant t. Furthermore, since this assumption holds only in abstraction, nothing prevents us from assuming that the mapping is smooth, that is, that the functions (1.2.1) are indefinitely differentiable. Thus, a change in shape of the body with time is determined by the change in distance between its points. Let the distance between two infinitely close points of the body ξ_i and $\xi_i + d\xi_i$ at the initial instant be ds_0, where

$$ds_0^2 = d\xi_1^2 + d\xi_2^2 + d\xi_3^2 \tag{1.2.2}$$

and at time t they occupy the positions x_i and $x_i + dx_i$. Then, the distance ds between them at time t can be determined by the relation

$$ds^2 = dx_1^2 + dx_2^2 + dx_3^2 \tag{1.2.3}$$

Relations of the type (1.2.2) and (1.2.3) are written as follows: If some expression is the sum of the products of the components of two (maybe identical) vectors a_i and b_i, we will simply write $a_i b_i$ and summation is implied. That is, if in any monomial the same subscript is repeated twice, it is to be assigned values of 1, 2, and 3 and the results obtained are to be summed (Einstein's summation convention). Thus, we get

$$a_i b_i = a_1 b_1 + a_2 b_2 + a_3 b_3 \tag{1.2.4}$$

Now (1.2.2) and (1.2.3) can be rewritten as

$$ds_0^2 = d\xi_i \, d\xi_i \quad \text{and} \quad ds^2 = dx_i \, dx_i \tag{1.2.5}$$

Furthermore, since the functions (1.2.1) are differentiable,

$$dx_i = \frac{\partial x_i}{\partial \xi_1} d\xi_1 + \frac{\partial x_i}{\partial \xi_2} d\xi_2 + \frac{\partial x_i}{\partial \xi_3} d\xi_3 = \frac{\partial x_i}{\partial \xi_k} d\xi_k$$

and $\tag{1.2.6}$

$$ds^2 = dx_i \, dx_i = \frac{\partial x_i}{\partial x_k} \frac{\partial x_i}{\partial x_l} d\xi_k \, d\xi_l$$

The change in distance between the two points of the body is characterized by the difference $ds^2 - ds_0^2$, that is,

$$ds^2 - ds_0^2 = \frac{\partial x_i}{\partial \xi_k} \frac{\partial x_i}{\partial \xi_l} d\xi_k \, d\xi_l - d\xi_i \, d\xi_i$$

$$= \left(\frac{\partial x_i}{\partial \xi_k} \frac{\partial x_i}{\partial \xi_l} - \delta_{kl} \right) d\xi_k \, d\xi_l \qquad (1.2.7)$$

where δ_{kl} is the Kronecker delta, which is equal to 1 for $k = l$ and equal to 0 for $k \neq l$. Now the change in the distance between the two infinitesimally close particles is totally described by the matrix in parentheses. Half of this matrix is called the strain tensor matrix with respect to the initial state and is denoted by ε_{kl}:

$$\varepsilon_{kl} = \frac{1}{2} \left(\frac{\partial x_i}{\partial \xi_k} \frac{\partial x_i}{\partial \xi_l} - \delta_{kl} \right) \qquad (1.2.8)$$

Let us now take into account the following circumstance. Since we assume that any particle of the body is identifiable at any instant, then knowing the position x_i of its center of mass at time t, we can find the corresponding position at the initial instant t_0. In other words equations (1.2.1) are solvable with respect to variables ξ_i:

$$\xi_i = \xi_i(\mathbf{x}, t) \qquad (1.2.9)$$

This implies that the Jacobian of (1.2.1) must not vanish at any time throughout the volume occupied by the body:

$$\det \left(\frac{\partial x_i}{\partial \xi_k} \right) \neq 0 \qquad (1.2.10)$$

From (1.2.9) we get

$$d\xi_i = \frac{\partial \xi_i}{\partial x_k} dx_k \qquad (1.2.11)$$

and

$$ds_0^2 = \frac{\partial \xi_i}{\partial x_k} \frac{\partial \xi_i}{\partial x_l} dx_k \, dx_l \qquad (1.2.12)$$

Now

$$ds^2 - ds_0^2 = \left(\delta_{kl} - \frac{\partial \xi_i}{\partial x_k} \frac{\partial \xi_i}{\partial x_l} \right) dx_k \, dx_l \qquad (1.2.13)$$

This relation differs from (1.2.7) in that the change in distance is now considered with respect to the current position x_i at time t rather than to the initial position ξ_i. Thus, we obtain the matrix of strain with respect to current state:

$$\varepsilon'_{kl} = \frac{1}{2}\left(\delta_{kl} - \frac{\partial \xi_i}{\partial x_k}\frac{\partial \xi_i}{\partial x_l}\right) \tag{1.2.14}$$

Both tensors (1.2.8) and (1.2.14) totally describe the change in the shape of the body.

Now we introduce the displacement vector **u** with components

$$u_i = x_i - \xi_i \tag{1.2.15}$$

Using (1.2.1) and (1.2.9), respectively, one can express the displacement in terms of ξ_i or x_i as

$$\begin{aligned}u_i(\boldsymbol{\xi}, t) &= x_i(\boldsymbol{\xi}, t) - \xi_i \\ u_i(\mathbf{x}, t) &= x_i - \xi_i(\mathbf{x}, t)\end{aligned} \tag{1.2.16}$$

whence

$$\begin{aligned}x_i(\boldsymbol{\xi}, t) &= u_i(\boldsymbol{\xi}, t) + \xi_i \\ \xi_i(\mathbf{x}, t) &= x_i - u_i(\mathbf{x}, t)\end{aligned} \tag{1.2.17}$$

and further

$$\frac{\partial x_i}{\partial \xi_k} = \frac{\partial u_i}{\partial \xi_k} + \delta_{ik}, \qquad \frac{\partial \xi_i}{\partial x_k} = \delta_{ik} - \frac{\partial u_i}{\partial x_k} \tag{1.2.18}$$

In what follows the partial derivatives of this form occur frequently, and it is convenient to introduce a more compact notation. Differentiation with respect to the initial coordinates is denoted by the usual convention; that is, a subscript (say, k) after a comma indicates the derivative with respect to the corresponding coordinate (ξ_k):

$$\frac{\partial x_i}{\partial \xi_k} = x_{i,k}, \qquad u_{m,l} = \frac{\partial u_m}{\partial \xi_l} \tag{1.2.19}$$

To denote differentiation with respect to current coordinates x_i we use a semicolon:

$$\xi_{i;k} = \frac{\partial \xi_i}{\partial x_k}, \qquad u_{i;k} = \frac{\partial u_i}{\partial x_k} \tag{1.2.20}$$

Now for strain components ε_{kl} we get

$$\varepsilon_{kl} = \tfrac{1}{2}(u_{k,l} + u_{l,k} + u_{i,k}u_{i,l}) \tag{1.2.21}$$

and

$$\varepsilon'_{kl} = \tfrac{1}{2}(u_{k;l} + u_{l;k} - u_{i;k}u_{i;l}) \tag{1.2.22}$$

In seismology, as a rule, linear theory is sufficient. In this case it is assumed that the gradients of displacement components $u_{i,k}$ are small and hence their products are negligible compared with themselves and, in turn, are negligible compared with unity. Then

$$\begin{aligned}
u_{i;k} &= \frac{\partial u_i}{\partial x_k} = \frac{\partial u_i}{\partial \xi_l}\frac{\partial \xi_l}{\partial x_k} \\
&= u_{i,l}(\delta_{kl} - u_{l;k}) \\
&= u_{i,k} - u_{i,l}u_{l;k} = u_{i,k}
\end{aligned}$$

That is, we need not distinguish between differentiation with respect to x_i and ξ_i. Then both expressions (1.2.21) and (1.2.22) have the same form:

$$\varepsilon_{kl} = \varepsilon'_{kl} = \tfrac{1}{2}(u_{k,l} + u_{l,k}) \tag{1.2.23}$$

Moreover, if it is assumed that the displacement components themselves are small, we can neglect them in expressions (1.2.17) and approximately put

$$x_i = \xi_i \tag{1.2.24}$$

That is, we do not distinguish the initial from the current coordinates. This corresponds to the transition to field theory, when only the initial configuration of the body is taken into account, and its motion is accounted for by assigning to every point of the initial configuration a value of the displacement vector. Here, in particular, the boundary of the body is considered stationary, and its change is accounted for by the displacement assigned to the points of it. In what follows, in place of ξ_i we will always write x_i and the notation $\partial u_i/\partial x_k = u_{i,k}$ will be retained for partial derivatives.

It is to be emphasized that the average strain and not the strain at a point has direct physical significance. Actually $u_{i,k}$ is the limit of the divided difference of the displacement

$$u_{i,k} = \lim_{|\Delta x|\to 0}\frac{\Delta u_i}{\Delta x_k}$$

where $|\Delta x|$ is the distance between the points of a body and Δu_i is their relative displacement. Since the particles considered to be physically infinitesimal must be sufficiently large, these limits should actually be understood as "asymptotic." When the distance between points is decreased, a range of distances (of the order of the size of a physically infinitesimal particle, the elementary scale) exists wherein the ratio $\Delta u_i / \Delta x_k$ does not change significantly, so that this change can be neglected. Then nothing prevents us from mentally permitting Δx to tend to zero. The local strain value thus obtained is then equal to the average strain over distances of the order of the elementary scale and depends on the choice of this scale.

Description of discontinuities within a continuous medium

So far, our efforts have been directed at understanding in what sense and to what degree the earth's material can be considered a continuum. We have sought some means to describe the motion of a continuous medium, and to this end have obtained the field of displacement vector and the field of strain tensor. It is clear that for our purposes we should not restrict ourselves to the description of continuous media since we are interested in the earthquake source, which, as we have already established, is a discontinuity in the earth's material. The presence of fracture implies that particles that are close together at one instant will be at a finite distance from each other at another instant. In that case, the concept of strain near the fracture surface loses its meaning because the mapping (1.2.1) is no longer smooth.

How should one proceed in this case? It is necessary to exclude fracture surfaces from the body; that is, those parts of the body that intersect the fracture surface ought not to be considered medium particles. In fact, to do this the entire layer adjoining the fracture surface with a thickness of the order of particle size should be excluded from the volume of the body. Thus, even if the fracture is a surface in a strict sense we have to consider it to be a layer of finite thickness. That is, within the accuracy of continuous description, the fracture has finite thickness and at the same time does not have any thickness since this thickness is considered infinitesimal. It is now possible to determine the strain for each point of this body that does not belong to the fracture, and the motion near the fracture (along the fracture) should be represented in some other way, such as by prescribing the relative positions of elements adjacent to both faces of the fracture. Evidently, it is sufficient to know the difference between the corresponding displacement vectors. Let us choose one face of the fracture surface $\Sigma(t)$ as positive and consider the normal n_i to this surface directed from the negative to the positive side. Then the motion on the surface is described by the displacement jump on it as

$$u_i^+ - u_i^- = a_i(\mathbf{x}, t) \qquad \text{at} \quad \Sigma(t) \tag{1.2.25}$$

where the symbols "+" and "−" refer to the values of displacement on different sides of the surface, and the point $\mathbf{x}$ belongs to the fracture surface. Thus, the kinematics of a body with a fracture can be described either by the field of displacement vector $u_i(\mathbf{x}, t)$ throughout the body or by the strain tensor $\varepsilon_{ik}(\mathbf{x}, t)$ outside the fracture and by the displacement jump $a_i(\mathbf{x}, t)$ at all points of the fracture. Here, if in the initial instant the fracture is not a "gaping fissure," then it is impossible for the sides of the fracture to come closer. This is expressed by the condition that the normal component of the displacement jump is positive:

$$a_{(n)} = a_i n_i \geq 0$$

In what follows, if an index is not subject to the summation convention, we will write it in parentheses, as in this equation.

This completes the description of the kinematics of a body with a fracture. However, it is not possible to apply the same considerations to all fractures in the medium in the case of earthquakes. The fact that earthquakes of different magnitudes occur, and that their energies differ by many orders, shows that fractures in the earth might be of many different sizes, ranging from a few or tens of meters to hundreds of kilometers. If we are dealing with earthquakes of a particular magnitude, we should take into account fractures of the corresponding scale, and other fractures of lesser size should be neglected. All the parts of the medium that might contain many smaller fractures are to be chosen as particles.

Then we can say not only that the rock sample with which we deal in the laboratory is not that continuous medium in which seismic waves propagate, but also that the medium in which large earthquakes occur is different from the medium in which smaller earthquakes occur.

One may argue that this is impossible since we are dealing with the same earth. Let us go back to the example of a spring for a better understanding of the nature of this contradiction. We have seen that the spring as a spring is something different from the spring as a piece of wire, and so on, and at the same time it is the same spring. In the same way, one and the same volume within the earth is not the same when it is considered as a continuum in which earthquakes occur, as a source of

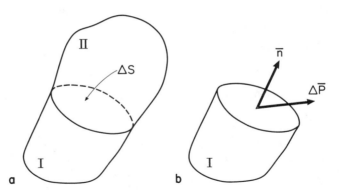

1.4   Definition of stress. See text for a description of parts a and b.

rock samples for laboratory experiments, or simply as some collection of molecules. Thus, these different media simply reflect the different qualitative aspects of one and the same earth. In other words, the earth represents different media with respect to different processes. Later we will, as a rule, ignore this fact, keeping in mind that the considerations of this section can be extended to every physical parameter and to every process involved. If need be, however, we will return to this question.

1.3   Dynamics of a continuous medium

Definition of stress

Cauchy introduced the concept of stress based on the following assumption: Two particles of a continuous medium interact through the surface separating them. More precisely, the effect of any particle of the medium on another particle can be replaced by forces acting on the surface of the second particle. Let us examine two particles of the medium (I and II in Figure 1.4a), separated by an area ΔS. It is postulated that the motion of particle I will not change if particle II is removed and some forces are applied to the area ΔS. The total force $\Delta \mathbf{P}$ and moment $\Delta \mathbf{M}$ of these forces depend on the magnitude and orientation of the area ΔS. It is assumed in classical mechanics of a continuous medium that the following limits exist,

$$\sigma(\mathbf{n}) = \lim_{\Delta S \to 0} \frac{\Delta \mathbf{P}(\Delta \mathbf{S})}{\Delta S}; \qquad \mathbf{m}(\mathbf{n}) = \lim_{\Delta S \to 0} \frac{\Delta \mathbf{M}(\Delta \mathbf{S})}{\Delta S} \qquad (1.3.1)$$

and the second of these vanishes. Here $\mathbf{n}$ is the unit external normal on the particle I boundary at ΔS and $\Delta \mathbf{S} = \mathbf{n} \, \Delta S$. Consider in what sense this assumption is to be understood according to the general phenomeno-logical approach to the continuous medium formulated in the previous section. To do so let us note that the size of particles and consequently the area ΔS cannot be made arbitrarily small without going outside the framework of the description of the material as a continuous medium. Consequently the limits (1.3.1) are to be taken as "asymptotic." That is, if the area ΔS is changed within certain limits, the ratio $\Delta \mathbf{P}/\Delta S$, known as the mean stress vector (traction), does not vary significantly, and hence this variation can be neglected. If we ignore it, it is easy to imagine ΔS tending to zero. It is clear that in such a case the traction $\sigma(\mathbf{n})$ depends on the scale of the area that is considered elementary. In other words, a multitude of different stresses corresponding to different levels of macroscopic description of this material as a continuous medium can be assigned to a given material. When speaking of "tectonic stresses" causing earthquakes, it is essential that we understand clearly which areas are considered elementary. Obviously, the area should be suffi-ciently large to ignore structural details (faults, blocks, etc.) that are "small" compared with the representative scale of the tectonic process under consideration. It is known that fractures at the earthquake source vary by many orders depending on the magnitude of the earthquake. However, it is to be understood that the description of a medium by the "smooth" model makes it possible to draw conclusions only on a given level of macroscopic description. For example, for a source considered to be a discontinuity in a smooth medium, the radiation characteristics of seismic waves are correct only in a range of sufficiently large wave-lengths. In other cases it is necessary to consider the possibility of a finer (microscopic) description of the medium and refer to "macrostresses" as well as "microstresses."

The traction $\sigma(\mathbf{n})$ depends linearly on the orientation of the area ΔS,

$$\sigma_i(\mathbf{n}) = \sigma_{ik} n_k \tag{1.3.2}$$

where the matrix σ_{ik} is called the stress tensor. Tensor σ_{ik} is symmetric:

$$\sigma_{ik} = \sigma_{ki} \tag{1.3.3}$$

In addition to the forces exerted directly by the neighboring particles, there are forces acting on a particle from distant parts of the body or from other bodies. In the case of earthquakes, these are gravitational forces that are proportional to the mass of the particles on which they act. Such forces are called body forces. The limit

$$\mathbf{f} = \lim_{\Delta m \to 0} \frac{\Delta \mathbf{F}}{\Delta m} \tag{1.3.4}$$

(where Δm is the mass of a particle and $\Delta \mathbf{F}$ the resultant body force acting on it) is called the body force density.

Equations of motion

Let us examine the motion of the center of mass of a part of the medium in a volume V with surface S. D'Alembert's principle states that the sum of all the forces acting on this part together with the forces of inertia should be equal to zero,

$$M\ddot{x}_{0i} + F_i = 0 \qquad (1.3.5)$$

where $\ddot{x}_{0i}$ is the acceleration of the center of mass, M the total mass of the volume, and F_i the total force on the volume. The acceleration of the center of mass can be expressed in the form

$$\ddot{x}_{0i} = \frac{1}{M} \int_V \ddot{x}_i \, dm = \frac{1}{M} \int_V \ddot{u}_i \, dm \qquad (1.3.6)$$

where $\ddot{u}_i$ is the second time derivative of the displacement vector. The total force F_i is represented as

$$F_i = \int_V f_i \, dm + \int_S \sigma_{ik} n_k \, dS \qquad (1.3.7)$$

where the first term is the total body force and the second the total surface force. Substituting (1.3.6) and (1.3.7) into (1.3.5) and keeping in mind that $dm = \rho \, dV$, where ρ is the density of the medium, we get

$$\int_V (f_i - \ddot{u}_i)\rho \, dV + \int_S \sigma_{ik} n_k \, dS = 0 \qquad (1.3.8)$$

Applying Green's theorem to this, we obtain

$$\int_V \left(\sigma_{ik,k} + \rho f_i - \rho \ddot{u}_i \right) dV = 0 \qquad (1.3.9)$$

This relation holds for an arbitrary volume V, not containing the discontinuities, and implies that the integrand should vanish everywhere (more correctly, almost everywhere) within the medium, that is,

$$\sigma_{ik,k} + \rho f_i = \rho \ddot{u}_i \qquad (1.3.10)$$

These equations are called the equations of motion.

Let us now examine the fracture surface Σ. We apply equation (1.3.8) to the cylindrical volume intersecting Σ, as shown in Figure 1.5:

$$\int_V (f_i - \ddot{u}_i)\rho \, dV + \int_{S_h} \sigma_{ik} n_k \, dS = -\int_{\Delta S^+} \sigma_{ik} n_k \, dS - \int_{\Delta S^-} \sigma_{ik} n_k \, dS \qquad (1.3.11)$$

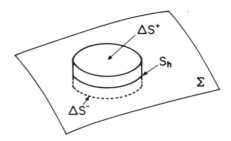

1.5   Continuity of stress across fracture surface.

Let the height h of this volume tend toward zero. The terms on the left side of equation (1.3.11) would also tend toward zero, and the terms on the right side would reduce to integrals over the portion ΔS of the fracture surface on both its sides, that is,

$$\int_{\Delta S} (\sigma_{ik}^+ - \sigma_{ik}^-) n_k \, dS = 0 \tag{1.3.12}$$

or, since ΔS is arbitrary,

$$\sigma_{ik}^+ n_k - \sigma_{ik}^- n_k = 0 \quad \text{on} \quad \Sigma(t) \tag{1.3.13}$$

Consequently, the stress vector is continuous across the fracture surface.

Hooke's law

The fifth and last assumption in the definition of the earthquake source states that the medium outside the fracture surface is elastic. This implies that the stress in the medium depends only on strain. In seismology it is necessary, however, to generalize this definition. Strain is only a relative measure of the change in shape of the particles of a body because it is measured with respect to a certain initial state. The so-called natural state of the body in which it is free from stress is usually taken as an initial state in the theory of elasticity. However, in formulating the theory of the earthquake source, the concept of such a "natural" state is meaningless because the rocks inside the earth are always under stress (at least hydrostatic – i.e., overburden pressure). It is also convenient to take the state of the medium at the moment just preceding the earthquake as the initial state from which the displacement and strain are measured. Then it must be assumed that there exists a one-to-one relation not between the strain and the total stress, but between the strain and the difference of stress in the current and initial states. Restricting ourselves to the linear theory of elasticity under the obvious assumption that the initial shear stress components are small compared with the elastic moduli, we get the linear dependence of the stress on strain,

$$\sigma_{ik} = \sigma_{ik}^0 + c_{iklm}\varepsilon_{lm} \tag{1.3.14}$$

where σ_{ik}^0 is the initial stress. It is assumed that the elastic potential exists, which implies the symmetry of the stiffness tensor:

$$c_{iklm} = c_{lmik} \tag{1.3.15}$$

Moreover, since the stress tensor is symmetric, c_{iklm} should be symmetric with respect to the transposition of indices in each pair, that is,

$$c_{iklm} = c_{kilm} \tag{1.3.16}$$

For the sake of simplicity, it is usually convenient to assume that the medium is isotropic. Then the stiffness tensor is expressed through Lamé's constants λ and μ as

$$c_{iklm} = \lambda \delta_{ik}\delta_{lm} + \mu(\delta_{il}\delta_{km} + \delta_{im}\delta_{kl}) \tag{1.3.17}$$

The medium is in equilibrium in the initial state; that is, the initial stresses must satisfy equations (1.3.10) without the inertia term:

$$\sigma_{ik,k}^0 + \rho f_i = 0 \tag{1.3.18}$$

It is assumed that the gravitation force f_i does not change during an earthquake. This is justified because even in very large earthquakes the displacement does not exceed a few meters. It will be convenient to introduce a special notation for stress perturbation during an earthquake as

$$\tau_{ik} = \sigma_{ik} - \sigma_{ik}^0 \tag{1.3.19}$$

Then subtracting equation (1.3.18) from the equation of motion (1.3.10), we get

$$\tau_{ik,k} = \rho \ddot{u}_i \tag{1.3.20}$$

and Hooke's law (1.3.14) can be expressed as

$$\tau_{ik} = c_{iklm}\varepsilon_{lm} \tag{1.3.21}$$

Equations (1.3.20) and (1.3.21) coincide in form with the usual equations of the theory of elasticity in the absence of body forces. At the same time, the expression for density of strain energy w_e has another form, which follows from (1.3.14),

$$w_e = \sigma_{ik}^0 \varepsilon_{ik} + \tfrac{1}{2}\tau_{ik}\varepsilon_{ik} \tag{1.3.22}$$

or

$$w_e = \tfrac{1}{2}(\sigma_{ik}^0 + \sigma_{ik})\varepsilon_{ik}$$

which does not coincide with the expression for the strain energy of an elastic medium relative to its natural state. This will be of significance when we consider earthquake energy.

Boundary conditions; description of fractures

Equation of motion (1.3.10) or (1.3.20), Hooke's law (1.3.14) or (1.3.21), and the definition of the infinitesimal strain tensor (1.2.23) comprise a complete system of differential equations governing the motion of a continuous linearly elastic medium. The initial and boundary conditions are to be added to these equations. At the initial instant $t = 0$ just before an earthquake, the medium is at rest; that is, the rate of displacement of all the particles is equal to zero. The displacement vector is measured from the state at $t = 0$; that is, it is equal to zero initially. This gives the initial condition as follows:

$$u_i = 0; \qquad \dot{u}_i = 0 \qquad \text{for} \quad t = 0 \tag{1.3.23}$$

The boundary conditions on the surface of the earth can be obtained if the boundary value of stress on it is equated to the atmospheric pressure on land and to the hydrostatic pressure of water at the bottom of the sea. These pressures remain unchanged during earthquakes. This leads to the condition

$$\tau_{ik} n_k = 0 \tag{1.3.24}$$

on the earth's surface. In addition to these conditions, certain conditions on the fracture surface at the source are also necessary. They can be formulated in various ways. One type of boundary condition on the fracture surface is given by equations (1.2.25) and (1.3.13). Since the initial traction is continuous across the fracture surface, the equation (1.3.13) can be rewritten in terms of the stress perturbations τ_{ik}. Thus, the boundary conditions on the fracture surface can be expressed as

$$\left.\begin{array}{l} \tau_{ik}^{+} n_k - \tau_{ik}^{-} n_k = 0 \\ u_i^{+} - u_i^{-} = a_i(\mathbf{x}, t) \end{array}\right\} \quad \text{in} \quad \Sigma(t) \tag{1.3.25}$$

This description of fracture will be called kinematic. It has several advantages. First of all, the time dependence of the fractured area is immaterial in this description. In fact, we can extend the fracture area arbitrarily beyond the actual fracture by assigning a vanishing displacement jump $a_i(\mathbf{x}, t)$ there. Furthermore, the principle of superposition is also valid for such a description. That is, the solution for several fractures described by kinematic conditions is simply the sum of the solutions for each of these fractures separately. A drawback of this description is that the displacement jump $a_i(\mathbf{x}, t)$ as a function of time and position on the fracture surface is not related to the initial stress state in the medium and/or the fault properties. Even if we accept, without reservation, that the normal displacement jump is zero, that is,

$$a_i(\mathbf{x}, t)n_i = 0 \quad \text{on} \quad \Sigma(t) \tag{1.3.26}$$

the boundary conditions on the fracture surface would be expressed in terms of two unknown functions of three variables (two spatial coordinates and time t).

This drawback can be partly avoided if, instead of assigning the relative motion $a_i(\mathbf{x}, t)$ of fracture faces, we formulate constitutive laws governing the interaction of these faces and the formation of the fractured area itself. The interaction of fracture faces is characterized by the forces that one face exerts on the other, that is, by the traction $\sigma_{ik}n_k$ on this surface.

The required constitutive laws will consist of material- and state-dependent relations between these traction components and the displacement jump vector and/or its rate. Remembering that the fracture surface may represent a band of localized shear deformation, we realize that normal (tensile) fracture is relatively independent of shear fracture in the sense that a surface broken in shear can conserve nonvanishing tensile strength. If the tractions at the surface have a one-to-one relation with the displacement jump, it is more convenient to denote it as an "elastic contact" or "joint" rather than as a "fracture," the derivatives of tractions by displacement jump components being called the "joint stiffness tensor." However, when this relation is not one-to-one, for example, if there is an irreversible displacement jump, the surface is called a "frictional contact" or "fracture" (crack). The behavior of a fracture relative to the normal displacement jump depends on the physical nature of the fracture. If there is an actual opening ("gaping fissure"), the faces of fracture do not interact, that is, $\sigma_{ik}n_k = 0$ and the surface (or its part) is said to be broken in tension. This is the simplest case and the most important in engineering applications. If the fracture is filled with some material or is macroscopically closed but contains open parts on the microscopic level, the presence of normal relative displacement does not necessarily imply the absence of interaction and usually is supposed to be related to tangential displacement jump (slip). The latter feature is called "joint dilatancy." When the faces of a surface interact but the slip is not in one-to-one relation to traction, the surface is said to be broken in shear. Once again, a surface experiencing shear fracture in one direction may conserve nonvanishing strength in the perpendicular direction (due, say, to "furrows" produced by previous sliding). Consequently, one and the same surface can be fractured in three independent modes. Usually, normal (tensile) fracture implies shear fracture, except in some very special cases like prefracturing of polymeric materials.

Thus, a sufficiently general form of the law governing a discontinuity is

$$\sigma_{ik} n_k = g_i\left(\mathbf{a}, \dot{\mathbf{a}}, \sigma_{(n)}, T, \kappa\right)$$

where g_i is some material function, T is the temperature, κ denotes all other possible relevant thermodynamic parameters, and $\sigma_{(n)} = \sigma_{ik} n_i n_k$.

A detailed analysis of this general form would result in the development of a general "rheology of discontinuities," including, in particular, friction and wear. Although extremely interesting, this subject is beyond the scope of this book. In what follows, we confine our considerations to the simplest case of fractures and/or conventional frictional contacts, which is probably sufficient for the theory of an earthquake in progress.

If the normal component of the displacement jump is nonzero (open fracture), the faces do not interact, that is, $\sigma_{ik} n_k = 0$, or, according to (1.3.19), $\tau_{ik} n_k = -\sigma_{ik}^0 n_k$. It would be reasonable to assume that the normal stress on the fracture surface cannot be tensile because it would give rise to opening of the fracture. If the normal stress $\sigma_{ik} n_i n_k \le 0$, that is, if it is compressive, interaction (friction) must occur between the fracture faces. This interaction is determined by the physical nature of the fracture, by the conditions on the fracture surface, and by the presence or absence of liquid or plastic phases, and so on. Unfortunately, there is little information on the laws of friction under the conditions prevailing at the earthquake source and hence some *a priori* laws of friction must be assumed based on the few experimental results that are available. In general, the frictional stress might depend on the normal stress, its direction, the slip rate $\dot{a}_i$, the temperature, and other thermodynamic parameters on the fracture surface. Hence, the "dynamic" boundary conditions on the fracture surface, when expressed in terms of stress, must have the form

$$\left. \begin{aligned} \sigma_{ik} n_k &= 0 && \text{when} \quad a_i n_i > 0 \\ a_i n_i &= 0 \\ \sigma_{(t)i} &= g_i\left(\dot{\mathbf{a}}, \sigma_{(n)}, T, \kappa\right) \end{aligned} \right\} \quad \text{when} \quad \sigma_{(n)} \ge 0 \qquad (1.3.27)$$

at $\Sigma(t)$, where

$$\sigma_{(n)} = \sigma_{ik} n_i n_k$$

$$\sigma_{(t)i} = \sigma_{ik} n_k - \sigma_{(n)} n_i$$

are the normal and shear tractions, respectively. In special cases some particular functions for g_i can be assumed. For the dry (Coulomb) friction law, which was shown by Byerlee (1967) to hold for rocks,

$$g_i = k\sigma_{(n)} \dot{a}_i (\dot{a}_m \dot{a}_m)^{-1/2} \tag{1.3.28}$$

where k is the coefficient of kinetic friction. If the fracture consists of a plastic slip band, then, neglecting work hardening, the governing law can be roughly expressed in a form close to (1.3.28),

$$g_i = \sigma_p(\sigma_{(n)}) \dot{a}_i (\dot{a}_m \dot{a}_m)^{-1/2} \tag{1.3.29}$$

where σ_p is the yield stress, which is assumed to be dependent on normal pressure. In the presence of a liquid phase, if the fracture consists of a band of molten material, for example, it can be assumed that the interaction of its faces can be described adequately by the law of viscous friction:

$$g_i = \sigma_\eta \dot{a}_i \tag{1.3.30}$$

where σ_η is the coefficient of viscous friction. The laws (1.3.28) and (1.3.29) can be further simplified if it is assumed that the occurrence of shear fracture has no significant effect on the magnitude of the normal stress, that is, if change in this stress is neglected in comparison with the magnitude of rock pressure. Such an assumption seems to be reasonable. Then (1.3.28) and (1.3.29) can be expressed in the form of the dry friction law:

$$g_i = g\dot{a}_i (\dot{a}_m \dot{a}_m)^{-1/2} \tag{1.3.31}$$

where the coefficient g depends only on the position on the fracture and, to a first approximation, can be taken as constant. It is convenient to express the conditions (1.3.28) in terms of the stress perturbation tensor τ_{ik} using the relation (1.3.19). Consider only the case of closed fracture, since tensile fracture is excluded by an assumption of Section 1.1. Then in place of (1.3.27) we get

$$\left.\begin{array}{c} \tau_{(t)i} = -\sigma_{(t)i}^0 + g_i \\ a_i n_i = 0 \end{array}\right\} \quad \text{on} \quad \Sigma(t) \tag{1.3.32}$$

where

$$\sigma_{(t)i}^0 = \sigma_{ik}^0 n_k - \sigma_{jk}^0 n_i n_j n_k; \qquad \tau_{(t)i} = \tau_{ik} n_k - \tau_{jk} n_i n_j n_k \qquad (1.3.33)$$

Let us, for the time being, assume that the fracture propagation history at the source [i.e., the position of the edge of $\Sigma(t)$] and the stress drop history on the fracture surface are known, that is,

$$\tau_{(t)i} = \tau_{(t)i}(\mathbf{x}, t) \qquad \text{at} \quad \Sigma(t) \qquad\qquad (1.3.34)$$

Equations (1.3.20) and (1.3.21), together with the initial conditions (1.3.23) and boundary conditions (1.3.24) and (1.3.32), constitute the linear elastic problem formulation. The uniqueness of its solution is subject to a condition at the fracture edge (edge condition). Thus, the displacement (in particular, seismic waves) due to the propagating fracture is determined [condition (1.3.32)] by the stress drop at the fracture surface and does not depend on the distribution of initial stress around it. However, the initial stress on the fracture surface is reflected in the stress drop and hence influences the earthquake source process.

Equations (1.3.27) through (1.3.34) are only the simplest examples. In particular, the above-mentioned joint dilatancy is not included in these equations. Some laws governing surface interaction incorporating the rate and state dependence of frictional sliding have been proposed on the basis of experimental data. Seismological research in this direction started with the experiments of Dieterich (1978, 1979), who measured the change in the shear stress across the cut surface of large rock samples against the slip across it for various sliding velocities at room temperature and on the order of tens of bars of normal stress. On the basis of these experiments, he postulated a constitutive law that accounts for slip history effects and can be written in the form

$$k = k_0 + A \log(Bt + 1)$$

where k was defined earlier, k_0, A, and B are constants, and t is the time of contact of the saw-cut surfaces. The transition of the frictional stress from the static to dynamic level was also studied by Scholz, Molnar, and Johnson (1972), Johnson and Scholz (1976), and Johnson (1981), among others, on smaller samples but at higher normal stresses. Using the results of Dieterich as well as further experimental data, Ruina (1980, 1983) formulated constitutive relations that depend on the slip rate and on one or more state variables. The general form of this constitutive relation can be written as

$$
\left.
\begin{array}{l}
S = f(\dot{a}, \theta_i) \\
\dot{\theta}_i = h_i(\dot{a}, \theta_i)
\end{array}
\right\} \quad i = 1, \ldots, n
$$

where S is the frictional strength of a surface undergoing unidirectional slip at a rate $\dot{a}$ and the θ_i's are the n-state variables, f and h_i being some functions obtained from the laboratory data. The direct application of such laws to the earthquake sliding process and the use of the results in interpreting earthquake data must always be made with some reserve, because laboratory samples with which such governing equations are obtained, whatever their size, are never representative on the earthquake scale and because of the obvious absence of long-term metamorphic and geochemical processes and microcracking at various scales.

2

GENERAL CONCEPTS OF FRACTURE MECHANICS

2.1 Introduction

The classical concepts of the mechanics of a continuous medium were sufficient for the formulation of the equations of motion and the boundary conditions. In order to formulate laws governing fracture propagation at the earthquake source, it is necessary to study the process of fracturing of the medium, that is, to consider the mechanics of brittle fracture.

Models of fracture

The term "fracture" will be used for the loss of continuity of a continuous medium, that is, the creation of discontinuities in a medium due to stress. Usually a distinction is made between a brittle and ductile fracture, a brittle fracture being called that which occurs without visible permanent deformation, and a ductile fracture being one that is accompanied by considerable inelastic deformation. This classification is convenient in engineering, where the result of fracture – the loss of inherent strength of a part or a structure – is of interest rather than the fracturing itself. For example, in a tensile test on a cylindrical sample the fracture is said to be brittle if there is no appreciable necking, and ductile otherwise. It is clear, however, that permanent deformation and fracture (breaking of the sample into parts) do not, in general, occur simultaneously, since the permanent deformation (in view of its definition) occurs in a continuous medium, that is, within the unfractured part of the material, and has no bearing on the process of fracture itself. For tensile testing, this means that the fracture occurs in the sample with the neck already present, when the plastic deformation is already complete. Consequently, if we wished to distinguish the concept of brittle fracture from that of ductile fracture more precisely, we would call a fracture brittle only if the material never underwent any irreversible deformation before a particular test. But this has an obvious implication: No material can fracture brittlely except, possibly, monocrystalline bodies with perfect, defectless structure. The rocks fractured at an earthquake source experience permanent inelastic tectonic deformations and hence this contraposition of ductile and brittle fractures can be applied to them even less. Consequently, Kostrov, Nikitin, and Flitman (1969) redefined brittle fracture as a fracture of the medium due to the formation of cracks. In this respect, the term is a synonym of "fracture" as it is used in the present book. Thus, the properties of the medium with respect to fracture generally do not depend on its properties with respect to the deformation (rheology), and one can consider brittle fracture of bodies with various rheologies – elastic, viscoelastic, plastic, and so on.

The fracture of the medium consists in the nucleation and propagation of cracks (fractures) or faults in this medium. Actual fracturing occurs at the edge of the crack during its propagation into a continuous medium. In fact, the particles of the medium adjacent to the crack faces are already fractured, so that the fracturing is complete there. But the particles ahead of the crack front remain continuous; that is, the fracturing has not commenced there. The behavior of particles in the continuous state is described by a rheological model of the medium, and that of fractured particles, those adjacent to the crack faces, is described by the friction law.

Generally, a third, intermediate state of particles can be introduced when the crack edge passes through the particle, and its behavior can be described neither by the rheological model nor by the friction law. On closer examination, this intermediate state can be nothing but continuous or discontinuous; that is, it must also be described either by a rheological model or by a friction law. The reason for considering some part of the medium or some part of the fracture surface (together usually called the "process zone") as a third state is that simplified rheological and frictional models fail to describe the behavior of material in the process zone, where it is unstable or close to instability. Consequently, the process zone is represented by either strain-dependent rheology or slip-dependent friction or by both.

The set of assumptions related to the transition of particles of the medium from the continuous state to the broken state has been called the "model of fracture" (Kostrov et al., 1969). The simplest model of fracture is obtained if it is assumed that there is no intermediate state for the particles of the medium. Here the details of the fracture process are totally excluded from consideration. Such a medium is called ideally or perfectly brittle. In an ideally brittle medium, particles pass from the continuous to the fractured state along a curve called the crack edge, the region where the particles are in the intermediate state being absent. Models of nonideally or imperfectly brittle bodies are obtained if one assumes that there exists some interaction between the crack faces near the edge, dependent on the displacement jump, which cannot be reduced to friction. The models suggested by Barenblatt (1959), Leonov and Panasyuk (1959), and Dugdale (1960) as well as the "cohesive zone" models used by Ida (1972) and by Palmer and Rice (1973) are models of nonideally brittle bodies.

The cohesive forces in nonideally brittle models are not necessarily interatomic ones. In fact, these forces are designated as tractions acting between the fault (crack) faces, and the very concept of stress is inseparably linked with the description of matter as a continuous medium and depends on the level of "macroscopicity" of the description. To make this discussion more specific, let us consider the physical phenomena accompanying the fracture. The crack edge in a metal plate with a thickness of around 1 mm is shown in Figure 2.1. In a region of 10^{-7} cm size, the material cannot be described by a continuous model, and the fracture (if it could be so called on this level) reduces simply to the rupture of interatomic bonds. On the scale of 10^{-6} cm the molecular structure of the material may be ignored, and the material can be roughly described as a continuous medium, preserving, however, the concept of elementary dislocations whose Burgers vectors are defined by the lattice constant. In the regions of 10^{-4} cm size it becomes meaningless and impossible to account for separate elementary dislocations that are merged into slip bands, and the subgrain interfaces must be taken into account. Next comes the effect of the polycrystalline structure of material and the description of the defects in terms of bulk plastic strain. In the regions of 10^{-1} cm and larger, it is possible to describe the material as a smooth continuous medium. Finally, in the region of about 100 cm, the whole sample can be considered a particle of the medium. Even these cases do not cover the entire range of scale levels involved in fracturing at the earthquake source. The roughest model is obtained by accounting for the faults ranging from a few kilometers to a few hundred kilometers (depending on the earthquake magnitude), faults of a smaller scale being smeared out by the choice of an elementary scale of, say, a few tens or hundreds of meters or more. Thus, the fracture of a continuous medium must be understood as fracture of that part of the material that is described as a continuous medium at a particular level of macroscopicity – that is, as the formation of discontinuities the size of

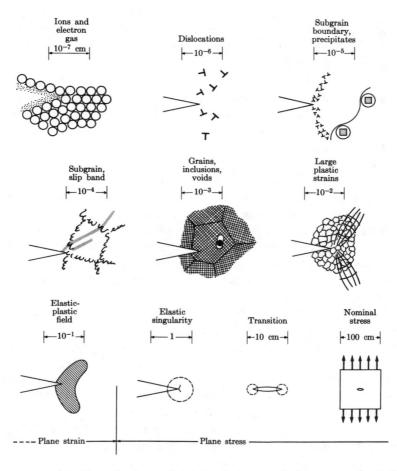

2.1   Example of a crack tip viewed at progressively coarser scales. (*After McClintock and Argon*, 1966. © *Addison-Wesley Publishing Co.*)

which exceeds the size of the particles of the material chosen as elementary. Hence, the formation of cracks of smaller size is not to be considered fracturing of the medium, although in a detailed (microscopic) examination of the fracture process the propagation of macroscopic discontinuities occurs by the nucleation and coalescence of these smaller "microscopic" fractures (microcracks). One of the greatest authorities on fracture mechanics, G. R. Irwin (1969), writes:

> The crack extension process is associated with growth and joining of advance origins and this tends to be discontinuous at fine scale. The speed of a real crack can, therefore, be defined and measured only as the average time rate of displacement of the apparent leading edge

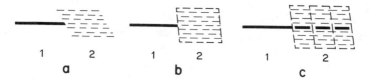

2.2 Smearing out of defects near the edge of a macrocrack. (a) Microfractures ahead of the main crack edge; (b) method of smearing out for an ideally brittle body; (c) method of smearing out for a model with cohesive forces. 1, Main fracture; 2, microfractures.

across distances much larger than the size of the small discontinuous units of separation. (p. 520)

This concept of the process of fracture growth is illustrated in Figure 2.2a. The smearing out of flaws ahead of the edge of a macroscopic crack can be performed in two ways. The first method consists in selecting a particle of the medium, where one is not interested in the process of fracturing because the particle is assumed to be physically infinitesimal (Figure 2.2b). Since the motion of this particle is described as the motion of its center of mass, microcracks totally disappear from consideration, and we obtain an ideally brittle model. However, the smearing out can be performed in another way (Figure 2.2c). Let us extend the fracture surface hypothetically and consider as particles only those parts of the body that do not intersect this surface. Then the displacement discontinuities across microcracks are averaged into some displacement jump across the extended surface. The average force of interaction between the faces is obtained by the summation of tractions acting on the continuous parts (on the "microscopic" scale) of the extended surface, and the result is taken as the cohesive force. In this way, a nonideally brittle model is obtained. The choice of one of these methods of smearing out is not entirely arbitrary. If the region with a significant density of microcracks extends over many "elementary" distances, the second method should be used. Otherwise, both methods of smearing out lead to the same result because the region of action of cohesive forces – Panasyuk's "zone of weakened bonds" – will be of the elementary size and must be considered a point.

For fractures of sufficiently large size, it is always possible to select a scale of macroscopic description for which the ideally brittle body model is valid. This means that, in the case of ideally brittle bodies, it is not possible to consider fracture of less than a definite size, which in turn is much larger than the elementary size. These smaller cracks must be smeared into the description of the material as a continuous medium. In particular, it follows from the above discussion that only the extension and not the nucleation of fractures can be described in an ideally brittle material. This is because the transition from "microscopic" to "macroscopic" fractures is discrete: Fractures of "submicroscopic" size (of the order of a few elementary sizes) would, in this case, be considered to be simultaneously existing (macroscopic) and nonexistent (microscopic). We thus obtain the notion of a material containing a multitude of defects (cracks, flaws, etc.) whose existence, though kept in mind, is nevertheless excluded from the continuous description. At the same time, any flaw whose growth leads to fracture, that is, to the formation of macroscopic discontinuity, should be considered macroscopic from the very beginning. In contrast, nonideally brittle models permit, generally speaking, the description of fracture nucleation as well.

Actually, the method of smearing out that leads to such a model can be used even if macrofracture is absent, if it is assumed that this fracture develops from the coalescence of microscopic fractures. In this case it is necessary to draw the surface of the future macrofracture through a concentration of these microfractures and apply the process of smearing out, as shown in Figure 2.2c. Then the entire fracture surface obtained will coincide with the zone of weakened bonds. Leonov and Panasyuk suggested their model simply to make the description of fracture nucleation possible. It is obvious from the previous discussion that even here microfractures are excluded from consideration, but a way is provided in which the initial defect from which a macrofracture originates can be considered as a point and the fracture development as continuous. But this initial defect, in a more detailed description of the medium, will be found to be a finite volume containing many "microfractures," which at this stage of description again have to be considered macroscopic. This in turn necessitates the consideration of a still lower level of description, and so on, until we reach the atomic or molecular level, where the concept of continuity of the medium and consequently that of fracture loses its meaning. Thus, the nonideally brittle models of fracture do not remove the qualitative difference between the macro- and microdefects; they only move it to finer scales.

The concept of multiple flaws in real material, the growth of the most dangerous of which leads to fracture, was proposed by Griffith (1921) to explain the fact that the strength of real materials is from two to three orders of magnitude lower than the theoretical strength implied by the perfect crystal structure of bodies. Griffith referred to the material considered to be a continuous medium at the level immediately above the discrete molecular structure. As a result, for a long time doubts were expressed as to the possibility of representing initial defects, several interatomic distances in size, by cracks in a continuous medium (Griffith cracks). Moreover, since Griffith's analysis referred to the submolecular level, it was not clear at the beginning how to apply these results to the fracture of real polycrystalline materials like steel. But actually Griffith's theory was not essentially related to the submolecular size of micro- cracks, and his technique could be simply extended to any macroscopic level by suitable redefinition of specific constants like the surface energy of material. This was done by Irwin (1948) and Orowan (1952), who extended Griffith's theory to microcracks in steel and proposed the concept of quasi-brittle fracture. The essence of this concept lies in considering fractures phenomenologically, including the region of irre- versible strains near a crack edge into the surface of fracture and introducing the concept of effective surface energy, which includes all the energy losses during fracturing and, in particular, plastic work. This made it possible to consider fractures in structural materials within the framework of the model of an ideally brittle body. The physical reason for such an interpretation is that the plastic zone around the crack edge is small (Figure 2.1). In earthquake source theory, the propagation of macrofracture is of primary interest, the elementary size being de- termined by the length of seismic waves recorded, ranging from a few meters to tens and even hundreds of kilometers. It can be expected that, under these conditions, the model of an ideally brittle body would describe the behavior of fractures fairly well.

Formal criteria of fracture

To formulate a quantitative description of fracturing, some relations linking the parameters of the model, that is, a formal fracture criterion, are necessary. The most widely accepted criterion is Griffith's criterion, which is simply the first law of thermodynamics formulated for any volume intersecting the edge of the fracture. Namely, to create a unit area of fracture surface, a specific amount of energy, called the effective specific surface energy, is necessary, which Griffith assumed to be a material constant. This criterion can be stated as follows:

$$\delta(A + Q) - \delta\Pi - 2\gamma\,\delta S = 0 \qquad\qquad (2.1.1)$$

Here $\delta(A + Q)$ is the work of external forces applied to a volume and the quantity of heat supplied to it, $\delta\Pi$ is the change of kinetic energy and the work of internal forces (the latter being the change in potential energy plus energy dissipation), δS is the fracture surface increment, and

γ is the effective specific surface energy, the factor 2 reflecting the two faces of the fracture surface. The terms "specific fracture work" and "crack-driving force" are also used to designate 2γ. Strictly speaking, the specific fracture work 2γ may depend on the velocity of fracture propagation, temperature, and other thermodynamic parameters; in other words, it is not a material constant but rather a material function of fracture velocity.

Griffith's criterion has a clear physical meaning but, being a global criterion, is not quite convenient for practical use. On the basis of Griffith's criterion (2.1.1), Irwin (1948) proposed a local fracture criterion for an ideally brittle elastic medium. To formulate Irwin's criterion, let us note that, in an ideally brittle elastic medium, neglecting the process zone size implies that the tractions ahead of the crack edge become infinite and vary as the inverse square root of the distance r from the edge:

$$\sigma_{ik} n_k = \frac{k_i}{\sqrt{2r}} \qquad (2.1.2)$$

where k_1, k_2, k_3 are called the stress intensity factors. Tensile and shear fracture modes are distinguished, the first also being called Mode I fracture. The shear mode, in turn, is split into Mode II, when the slip vector is normal to the crack edge (inplane shear), and Mode III, when the slip vector is tangent to it (antiplane shear). The coordinate system can be chosen such that the stress intensity factor k_1 corresponds to Mode I, whereas k_2 and k_3 correspond to Mode II and Mode III, respectively. In engineering, the tensile mode is of major importance, and in most cases the so-called condition of local symmetry, $k_2 = k_3 = 0$, is fulfilled. For this case, Irwin showed that, in the equilibrium state, when an increase in the load results in an extension of the fracture, the stress intensity factor has a critical value k_{1c}, which characterizes the fracture toughness. Thus, for tensile fracture, Irwin's criterion is

$$k_1 = k_{1c} \qquad (2.1.3)$$

For plane strain, the critical stress intensity factor is related to the specific fracture work by

$$k_{1c}^2 = \frac{2\gamma E}{\pi(1 - \nu^2)} \qquad (2.1.4)$$

where E is Young's modulus and ν is Poisson's ratio. For a propagating crack, the fracture criterion (2.1.3) may also be adopted, k_{1c} being, however, dependent on the fracture velocity.

Let us briefly describe some fracture criteria proposed for models of nonideally brittle bodies. Barenblatt's (1959) model is defined by the following assumptions: (1) Cohesive forces exist between the faces of the fracture surface at its edge; (2) the width of the edge zone where the cohesive forces act is much smaller than the crack dimensions and all the other dimensions of the body; (3) in the state of equilibrium, the distribution of cohesive forces and of the opening displacement is autonomous, that is, determined only by the properties of the material. In this model, finiteness of stress around the crack edge is used as the fracture criterion. For an elastic medium, this criterion reduces (for Mode I) to:

$$k_1 = \int_0^d g(s)\frac{ds}{\sqrt{s}} = \frac{N}{\pi} \tag{2.1.5}$$

where k_1 is the stress intensity factor, calculated neglecting the cohesive forces, $g(s)$ is the distribution of cohesive forces versus distance s from the crack edge, and d is the width of the edge zone, that is, $g(s) = 0$ for $s > d$. The constant N was termed the modulus of cohesion by Barenblatt. It is quite clear that Barenblatt's formal criterion (2.1.5) coincides with Irwin's criterion (2.1.3), except for notation. As a result, in crack problems for elastic media, Barenblatt's model is equivalent to the ideally brittle model with Irwin's criterion.

As a second example, let us consider the Leonov–Panasyuk model, in which it is assumed that constant cohesive stress σ_0 acts between the fracture faces in the zone of weakened bonds. In this case, two criteria determining the positions of two edges of the zone of weakened bonds, namely, where the fracture process begins and where it is completed, are required. As the first criterion, the condition of stress finiteness at the crack edge,

$$k_1 = 0 \tag{2.1.6}$$

is assumed. As the second criterion, it is assumed that at a certain crack-opening displacement δ, a material constant, the cohesive forces vanish – that is, at the transition from the zone of weakened bonds to that of broken bonds:

$$\left(u_i^+ - u_i^-\right)n_i = \delta \tag{2.1.7}$$

In this model, the specific fracture work is not a material constant, and consequently Griffith's criterion (2.1.1) is not equivalent to conditions (2.1.6) and (2.1.7). The reason is that, owing to a redistribution of the displacement jump, additional energy may be consumed or even released from the edge zone.

The "slip-weakening" model of Palmer and Rice (1973) for nonideally brittle bodies has much in common with the cohesive force models described earlier. In particular, the strength of the material in the process zone is assumed to vary with slip from some peak value, say σ^{stat}, to some residual frictional value, say σ^{kin}, when the slip exceeds some critical value, say a_0. The specific fracture work in this case is given by the integral (Rice, 1980)

$$\int_0^{a_0} \left(\sigma^{\text{stat}}(a) - \sigma^{\text{kin}} \right) da$$

When the relation between the stress and slip in the process zone is linear, this model is called the linear slip-weakening model. We shall discuss the application of this model to dynamic fracture problems in Chapter 5.

2.2 Energy conservation law

In what follows we shall consider only an ideally brittle medium. For a nonideally brittle medium, the same considerations apply (Kostrov, Nikitin, and Flitman, 1969).

For any body (or part of a body) the energy conservation law can be written as

$$\dot{A} + \dot{Q} - \dot{U} - \dot{K} = 0 \tag{2.2.1}$$

where $\dot{A}$ is the rate of work of external forces, $\dot{Q}$ the rate of heat supplied to the body (heat power), and $\dot{U}$ and $\dot{K}$ the rates of change of the internal and kinetic energies, respectively.

Let us first consider some volume V, not containing the fracture surface. The rate of work of external forces can be expressed as

$$\dot{A} = \int_S \sigma_{ij} n_j \dot{u}_i \, dS + \int_V \rho f_i \dot{u}_i \, dV \tag{2.2.2}$$

where n_j is the external unit normal to the surface S, enclosing volume V. For the heat rate $\dot{Q}$ the following expression holds:

$$\dot{Q} = - \int_S q_i n_i \, dS + \int_V q \, dV \tag{2.2.3}$$

where q_i is the heat flux vector from the body and q the specific heat production (e.g., due to radioactive decay). From now on, internal sources of heat will be neglected ($q = 0$). The rates of the internal and kinetic energies are

$$\dot{U} + \dot{K} = \frac{d}{dt}\left[\int_V(e + \tfrac{1}{2}\dot{u}_i\dot{u}_i)\rho\, dV\right] \tag{2.2.4}$$

where ρe is the density of internal energy.

We substitute equations (2.2.2) through (2.2.4) into (2.2.1) and integrate by parts. Taking into account that the volume V does not depend on time, time differentiation can be performed under the integral sign, which gives

$$\int_V\left\{\rho[\dot{e} + \dot{u}_i\ddot{u}_i - f_i\dot{u}_i] - (\sigma_{ij}\dot{u}_i)_{,j} + q_{i,i}\right\} dV \tag{2.2.5}$$

Consider the second term in braces. Differentiating, we get

$$(\sigma_{ij}\dot{u}_i)_{,j} = \sigma_{ij,j}\dot{u}_i + \sigma_{ij}\dot{u}_{i,j} \tag{2.2.6}$$

Taking into account the symmetry of the stress tensor, the equation of motion (1.3.10), and the strain definition (1.2.23), we obtain

$$(\sigma_{ij}\dot{u}_i)_{,j} = \rho\ddot{u}_i\dot{u}_i - \rho f_i\dot{u}_i + \sigma_{ij}\dot{\varepsilon}_{ij} \tag{2.2.7}$$

We substitute this expression into equation (2.2.5) to obtain

$$\int_V(\rho\dot{e} - \sigma_{ij}\dot{\varepsilon}_{ij} + q_{i,i})\, dV = 0 \tag{2.2.8}$$

This relation holds for any volume that does not contain a fracture. Owing to the arbitrariness of the volume, the integrand must vanish identically, that is,

$$\rho\dot{e} = \sigma_{ij}\dot{\varepsilon}_{ij} - q_{i,i} \tag{2.2.9}$$

This is the local form of the energy conservation law for a continuous medium. The first term on the right side of (2.2.9) is sometimes called the rate of internal work.

Now let the volume V intersect the fracture surface $\Sigma(t)$ over an area $\Delta\Sigma$ not containing its edge. The equations (2.2.2) and (2.2.3) are still applicable, but integration by parts is not applicable to the first term on the right sides of (2.2.2) and (2.2.3) because the integrands are discontinuous across $\Delta\Sigma$. To overcome this difficulty, we divide the volume V into two parts separated by $\Delta\Sigma$. Now integration by parts is permissible for both subvolumes, which gives

$$\int_{S^\pm}\sigma_{ij}n_j\dot{u}_i\, dS = \pm\int_{\Delta\Sigma}\sigma_{ij}n_j\dot{u}_i\, dS + \int_{V^\pm}(\sigma_{ij}\dot{u}_i)_{,j}\, dV \tag{2.2.10}$$

where $S^\pm$ are the parts of surface S containing parts $V^\pm$ of volume V on the positive and negative sides of $\Delta\Sigma$, respectively. Adding these expressions and using equation (1.3.13), we obtain

$$\int_S \sigma_{ij} n_j \dot{u}_i \, dS = \int_{\Delta\Sigma} \sigma_{ij} \dot{a}_i n_j \, dS + \int_V (\sigma_{ij} \dot{u}_i)_{,j} \, dV \qquad (2.2.11)$$

where, as before, a_i is the displacement jump at $\Delta\Sigma$. Similarly,

$$\int_S q_i n_i \, dS = \int_{\Delta\Sigma} (q_i^+ - q_i^-) n_i \, dS + \int_V q_{i,i} \, dV \qquad (2.2.12)$$

An additional term should be added to expression (2.2.4), which corresponds to the internal energy of the surface $\Delta\Sigma$. This gives

$$\dot{U} + \dot{K} = \frac{d}{dt}\left[\int_V (e + \tfrac{1}{2}\dot{u}_i\dot{u}_i)\rho \, dV + 2\int_{\Delta\Sigma} \gamma \, dS\right] \qquad (2.2.13)$$

where γ is the specific internal surface energy.

We now substitute (2.2.11) through (2.2.13) into (2.2.1) and transform as before, obtaining instead of (2.2.8)

$$\int_V (q_{i,i} - \sigma_{ij}\dot{\varepsilon}_{ij} + \rho\dot{e}) \, dV = \int_{\Delta\Sigma}\left\{(\sigma_{ij}\dot{a}_i - \Delta q_i)n_i - 2\dot{\gamma}\right\} dS \qquad (2.2.14)$$

where

$$\Delta q_i = q_i^+ - q_i^-$$

Obviously,

$$\left.\begin{array}{r} \Delta q_i n_i = \Delta q \\[2mm] \sigma_{ij} n_j = g_i \end{array}\right\} \quad \text{at} \quad \Sigma(t) \qquad (2.2.15)$$

where Δq is the heat flux from unit area of $\Delta\Sigma$ and g_i was defined in Section 1.3 as the frictional traction. The left side of (2.2.14) vanishes in view of (2.2.9). Consequently, the right side should also vanish. Then, owing to the arbitrariness of $\Delta\Sigma$, we get

$$g_i \dot{a}_i = 2\dot{\gamma} + \Delta q \quad \text{at} \quad \Sigma(t) \qquad (2.2.16)$$

This relation implies that the frictional work is partly spent in the change of internal surface energy and partly released into the medium as heat. Let us assume that the surface energy depends only on the thermodynamic state of the fracture surface itself and not on the relative motion of its faces. If, as an example, the thermodynamic state is determined by the surface temperature, that is, $\gamma = \gamma(T)$, then equation (2.2.16) can be rewritten as

$$g_i \dot{a}_i = 2\frac{\partial \gamma}{\partial T}\dot{T} + \Delta q \qquad (2.2.17)$$

It is appropriate to make a few observations on the concept of heat rate $\dot{Q}$. At the molecular level of the description of matter, it is impossible to define the concept of heat supplied to a body since the energy exchange of a system with its environment results from the direct interaction of molecules and must be considered to be the effect of external forces on molecules belonging to the system. In a macroscopic description of a set of molecules as a continuous medium, the portion of the energy supplied by the environment would be described as the work of tractions acting on the surface of the body. The portion of the energy that cannot be so described is called the quantity of heat supplied to the body. Now, according to our general phenomenological approach to a continuous medium given in Chapter 1, we must consider as heat the energy supplied to the body in any way that does not reduce to the work of external tractions. But what we call stress depends on the size of the particles of the medium that we consider infinitesimal. Then rapidly varying stresses at the previous scale level do not affect the magnitude of stress at the macrolevel. At this level, their work cannot be described mechanically; it must be included in the heat term. In particular, this is true for the energy transferred by sufficiently short waves that were smeared out in the macroscopic description of the medium (radiation losses). Thus, in (2.2.16) and (2.2.17), Δq contains not only the actual heat Δq_T but also the radiation loss Δq_r associated with the radiation of short elastic waves that are not taken into account by the model of the medium. The difference between heat and radiation loss should not be overemphasized because, due to damping, the short-wave radiation is ultimately transformed into heat at some distance, and the shorter the wavelength the shorter is this distance.

Let us now consider the volume V intersecting the fracture surface and containing part of its edge. In this case, in the expression (2.2.13) we cannot differentiate with respect to time under the integral sign because the integrand might be (and must be, as it will be seen later) singular at the fracture edge and the resulting integral might be divergent. Therefore, let us exclude a small toroidal volume V_ε of radius ε from the volume V, which contains the part $\Delta L(t)$ of the fracture edge $L(t)$, which intersects the volume V. Obviously, the volume V_ε propagates with the edge $\Delta L(t)$. We shall denote the remaining part of the volume by

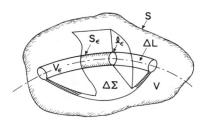

2.3   Region of integration at the propagating fracture edge.

$V - V_\varepsilon$ (Figure 2.3). Now the integral over the volume $V - V_\varepsilon$ can be differentiated under the integral sign, and letting the volume V_ε tend to zero, we obtain

$$\dot{U} + \dot{K} = \lim_{\varepsilon \to 0} \frac{d}{dt} \left[\int_{V - V_\varepsilon} \left(e + \tfrac{1}{2} \dot{u}_i \dot{u}_i \right) \rho \, dV + 2 \int_{\Delta\Sigma(t)} \gamma \, dS \right] \qquad (2.2.18)$$

We perform the differentiation taking into account that the volume $V - V_\varepsilon$ and surface $\Delta\Sigma(t)$ depend on time and obtain

$$\dot{U} + \dot{K} = \lim_{\varepsilon \to 0} \left[\int_{V - V_\varepsilon} \left(\dot{e} + \dot{u}_i \ddot{u}_i \right) \rho \, dV + 2 \int_{\Delta\Sigma(t)} \dot{\gamma} \, dS \right.$$
$$\left. + \int_{S_\varepsilon} \left(e + \tfrac{1}{2} \dot{u}_i \dot{u}_i \right) \rho v_j n_j \, dS + 2 \int_{\Delta L(t)} \gamma v \, dl \right] \qquad (2.2.19)$$

where v_j is the velocity of the fracture edge and $v = \sqrt{v_j v_j}$ its magnitude. Similarly differentiating expressions (2.2.11) and (2.2.12), we obtain

$$\dot{A} + \dot{Q} = \int_{\Delta\Sigma(t)} \left(\sigma_{ij} n_j \dot{a}_i - \Delta q \right) dS$$
$$+ \lim_{\varepsilon \to 0} \left[\int_{V - V_\varepsilon} \left[\left(\sigma_{ij} \dot{u}_i \right)_{,j} - q_{i,i} + \rho f_i \dot{u}_i \right] dV \right.$$
$$\left. - \int_{S_\varepsilon} \left(\sigma_{ij} \dot{u}_j - q_i \right) n_i \, dS \right] \qquad (2.2.20)$$

Substituting into (2.2.1), we obtain

$$\dot{A} + \dot{Q} - \dot{U} - \dot{K} = \int_{\Delta\Sigma(t)} \left[\sigma_{ij} n_j \dot{a}_i - \Delta q - 2\dot{\gamma} \right] dS$$

$$+ \lim_{\varepsilon \to 0} \int_{V-V_\varepsilon} \left[(\sigma_{ij} \dot{u}_i)_{,j} - q_{i,i} - \rho\dot{e} \right.$$

$$- \rho\dot{u}_i\ddot{u}_i + \rho f_i \dot{u}_i \Big] dV$$

$$- \lim_{\varepsilon \to 0} \int_{S_\varepsilon} \left[\sigma_{ij}\dot{u}_j - q_i + \rho\left(e + \tfrac{1}{2}\dot{u}_j\dot{u}_j \right) v_i \right] n_i \, dS$$

$$- 2 \int_{\Delta L(t)} \gamma v \, dl = 0 \qquad\qquad (2.2.21)$$

In view of equations (2.2.7), (2.2.9), and (2.2.16), the first two terms vanish and we obtain

$$2 \int_{\Delta L(t)} \gamma v \, dl + \lim_{\varepsilon \to 0} \int_{S_\varepsilon} \left[\sigma_{ij}\dot{u}_j - q_i + \rho\left(e + \tfrac{1}{2}\dot{u}_j\dot{u}_j \right) v_i \right] n_i \, dS = 0$$

$$(2.2.22)$$

Denote by l_ε the curve obtained as the cross section of surface S_ε, normal to the fracture edge. Then the last equation can be rewritten as

$$\int_{\Delta L(t)} \left\{ 2\gamma v + \lim_{\varepsilon \to 0} \int_{l_\varepsilon} \left[\sigma_{ij}\dot{u}_j - q_i + \rho\left(e + \tfrac{1}{2}\dot{u}_j\dot{u}_j \right) v_i \right] n_i \, dl \right\} dL = 0$$

$$(2.2.23)$$

Owing to the arbitrariness of the volume V and, consequently, of $\Delta L(t)$, the integrand must vanish, and

$$2\gamma v + \lim_{\varepsilon \to 0} \int_{l_\varepsilon} \left[\sigma_{ij}\dot{u}_j - q_i + \rho\left(e + \tfrac{1}{2}\dot{u}_j\dot{u}_j \right) v_i \right] n_i \, dl = 0 \quad (2.2.24)$$

Since, in the linear theory, density ρ does not depend on time, integration of equation (2.2.9) with respect to time gives

$$\rho e = \rho e_0 + \int_0^t (\sigma_{ij}\dot{\varepsilon}_{ij} - q_{i,i}) \, dt \qquad\qquad (2.2.25)$$

where e_0 is the density of internal energy at $t = 0$. We substitute this into (2.2.24), taking into account that e_0 is finite at the fracture edge.

Then, the corresponding term will vanish in the limit, so that finally

$$2\gamma v + \lim_{\varepsilon \to 0} \int_{l_\varepsilon} \left\{ \sigma_{ij} \dot{u}_j + v_i \left[\int_0^t \sigma_{kj} \dot{\varepsilon}_{kj} \, dt + \tfrac{1}{2} \rho \dot{u}_j \dot{u}_j \right] \right\} n_i \, dl$$

$$- \lim_{\varepsilon \to 0} \int_{l_\varepsilon} \left\{ q_i + v_i \int_0^t q_{j,j} \, dt \right\} n_i \, dl = 0 \qquad (2.2.26)$$

Here n_i is a unit vector, normal to l_ε and to $L(t)$ and directed into the medium. Thus, the last term represents the heat flux from the fracture edge and the second, taken with the opposite sign, represents the mechanical energy flux, that is, work done on the fracture edge.

2.3   Conditions of dynamic fracture onset and arrest

Propagation condition
The relation (2.2.26) expresses the law of energy conservation at the fracture edge: Mechanical work is partly converted to surface energy and partly dissipated in the form of heat. We denote

$$\gamma_{\text{eff}} = \gamma - \frac{1}{2v} \lim_{\varepsilon \to 0} \int_{l_\varepsilon} \left\{ q_i + v_i \int_0^t q_{j,j} \, dt \right\} n_i \, dl \qquad (2.3.1)$$

This expression includes the increase in surface energy per unit area of fracture as well as the heat loss per unit area due to irreversible processes during fracture. Following Irwin (1948) this quantity, which expresses the total loss of mechanical energy during fracture, will be called the effective surface energy. Now the relation (2.2.26) can be rewritten as

$$\gamma_{\text{eff}} = -\frac{1}{2v} \lim_{\varepsilon \to 0} \int_{l_\varepsilon} \left\{ \sigma_{ij} \dot{u}_j + v_i \left[\int_0^t \sigma_{jk} \dot{\varepsilon}_{jk} \, dt + \tfrac{1}{2} \rho \dot{u}_j \dot{u}_j \right] \right\} n_i \, dl \qquad (2.3.2)$$

Let the fracture edge $L(t)$ in the fixed system of coordinates $Ox_1x_2x_3$ be given by the equations

$$x_i = x_i^0(t, s) \qquad (2.3.3)$$

where s is the parameter determining the position of the point on the edge. To analyze expression (2.3.2) we will introduce the moving system of coordinates ξ_i, with its origin at a point on the edge (Figure 2.4). Let us take the ξ_1 axis along the normal to the fracture surface, the ξ_2 axis along the normal to the edge in the plane tangent to the fracture surface (i.e., collinear with v_i), and the ξ_3 axis along the edge. Let the fixed

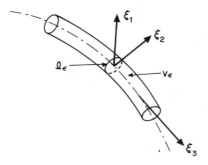

2.4   Moving coordinate system at the crack edge.

system of coordinates coincide with the moving system at the instant t. Then for instants close to t we have

$$\xi_1 = x_1; \qquad \xi_2 = x_2 - x_2^0(t); \qquad \xi_3 = x_3 \qquad (2.3.4)$$

According to Griffith's concept, the work spent in fracture must not vanish. Consequently, the limit in equation (2.3.2) also must not vanish. Therefore, as $\xi_i \to 0$, the integrand must have a singularity of the order r^{-1}, where $r = (\xi_1^2 + \xi_2^2)^{1/2}$. This means that at least some components of stress and particle velocity near the fracture edge must be singular. It is obvious that the integral with respect to time in (2.3.2) must have a singularity of the order r^{-1} if its contribution to the fracture work is finite. Then its integrand must be of the order of r^{-2} as $\xi_i \to 0$.

Differentiation and integration operations with respect to time in (2.3.2) are carried out in the fixed frame of reference x_i and are applied to the functions having a singularity at the moving point $\xi_i = 0$. In that case, the operation with respect to time for the leading term of asymptotic expansions can be replaced by operations with respect to the coordinate ξ_2. Let the function $f(x_1, x_2, t)$ have the form

$$f(x_1, x_2, t) = r^{-\kappa} f_0(\xi_1, \xi_2(t), t) \qquad (2.3.5)$$

where $\kappa > 0$ and f_0 is finite and its spatial derivatives increase more slowly than r^{-1} as $r \to 0$. Then it is obvious that

$$\left. \frac{\partial f}{\partial t} \right]_{x_i = \text{const}} = \kappa v \xi_2 r^{-\kappa - 2} f_0(\xi_1, \xi_2, t) + O(r^{-\kappa - 1})$$

$$\left. \frac{\partial f}{\partial \xi_2} \right]_{t = \text{const}} = -\kappa \xi_2 r^{-\kappa - 2} f_0(\xi_1, \xi_2, t) + O(r^{-\kappa - 1})$$

$$(2.3.6)$$

It is seen that differentiation with respect to time and ξ_2 increases the order of singularity by unity and, moreover, the asymptotic relation

$$\frac{\partial f}{\partial t}\bigg]_{x_i=\text{const}} \approx -v\frac{\partial f}{\partial \xi_2}\bigg]_{t=\text{const}} \tag{2.3.7}$$

holds. Now, let us assume that an integral with respect to time has the same properties as f in (2.3.5). Then it is easy to prove the following asymptotic relation:

$$\int^t \phi(x_1, x_2, t)\, dt = -\frac{1}{v}\int^{\xi_2} \phi(\xi_1, r + x_2^0(t), t)\, dr \tag{2.3.8}$$

Here, the lower limit can be chosen arbitrarily, including the value $t = -\infty$ and $\xi_2 = \infty$ subject to the convergence of the integrals.

Since the contour of integration in (2.3.2) tends to the point $\xi_i = 0$, the asymptotic relations (2.3.8) can be used. Then the expression (2.3.2) takes the form

$$\gamma_{\text{eff}} = \lim_{\varepsilon \to 0} \frac{1}{2}\int_{l_\varepsilon}\left\{\sigma_{ij}u_{j,2} - \delta_{i2}\left[\tfrac{1}{2}\rho v^2 u_{j,2}u_{j,2} + \int^{\xi_2}\sigma_{jk}\varepsilon_{jk,2}\, d\xi_2\right]\right\}n_i\, dl \tag{2.3.9}$$

where $u_{j,2}$ is the derivative of u_j with respect to ξ_2.

As observed earlier, only the singular part of the integral expression (2.3.9) contributes to γ_{eff}. The singular part is not uniquely defined. We specify the choice of singular parts of the components of stress and particle velocity, requiring that they satisfy not only asymptotically, but exactly, the relations in (2.3.7). Then the singular part will satisfy the equation

$$\sigma'_{ij,j} = \rho v^2 u'_{i,22} \tag{2.3.10}$$

obtained from the equations of motion (1.3.10) using (2.3.7), where primes denote the singular solution. Here the body forces f_i are omitted because they are finite and do not contribute to the singular solution. By substitution of the singular part thus chosen into (2.3.9), the resulting integral can be proved to be independent of the contour of integration l_ε. In fact, it can be shown that the two-dimensional divergence of the expression within braces vanishes.

Moreover, in the expression obtained, instead of the total stress σ_{jk}, change of stress τ_{jk} relative to the initial state (from which the displacement and strain are measured – in our case, the time of the earthquake onset) can be used, because the initial stress is continuous at $\xi_j = 0$.

Thus, the expression (2.3.9) for the specific fracture work can be written in the form

$$\gamma_{eff} = \frac{1}{2} \int_l \left\{ \tau'_{ij} u'_{j,2} - \delta_{i2} \left[\frac{1}{2} \rho v^2 u'_{j,2} u'_{j,2} + \int^{\xi_2} \tau'_{jk} \varepsilon'_{jk,2} \, d\xi_2 \right] \right\} n_i \, dl \qquad (2.3.11)$$

where τ'_{ij}, u'_j indicate the singular solution satisfying the asymptotic equation of motion

$$\tau'_{ij,j} = \rho v^2 u'_{i,22} \qquad (2.3.12)$$

and the contour l is an arbitrary curve surrounding the fracture edge and lying in the plane normal to the edge ($\xi_1 \xi_2$ plane). In the quasi-static case, (2.3.11) reduces to the well-known Γ integral (Cherepanov, 1967, 1979) or J integral (Rice, 1968). The independence of the expression (2.3.9) or (2.3.11) from the choice of contour l is important since only under this condition can the effective surface energy γ_{eff} be considered a well-defined physical quantity. Moreover, this makes possible further simplification of the expression (2.3.11). For this, let us choose the contour l in the form of a rectangle of width 2ε and arbitrary length $\xi''_2 - \xi'_2$, but such that the interval (ξ'_2, ξ''_2) contains the point $\xi_2 = 0$. Now if we reduce ε to zero, the integrals along the vertical sides will vanish. On the horizontal sides, the only nonvanishing component n_i is n_1, that is, $\delta_{i2} n_i = 0$, and we will obtain

$$\gamma_{eff} = - \lim_{\varepsilon \to 0} \int_{\xi'_2}^{\xi''_2} (\tau'_{i1} u'_{i,2}) \Big]_{\xi_1 = -\varepsilon}^{\xi_1 = +\varepsilon} d\xi_2 \qquad (2.3.13)$$

Velocity dependence of γ_{eff}; dynamics and kinematics of fractures

In deriving equations (2.3.9) through (2.3.13), we made no assumptions regarding the rheological properties of the medium. Actually, we assumed only continuity of the medium and Griffith's hypothesis that some work is required to create a fracture surface. If we add to these hypotheses the assumption that γ_{eff} is a material constant, we obtain a local form of Griffith's criterion for a perfectly brittle medium with arbitrary rheology. However, according to the definition (2.3.1), γ_{eff} contains not only the "true" surface energy but also the losses in the process of fracture expressed by the heat flow from the fracture edge. Moreover, the same reasoning is applicable to the "true" surface energy itself as to stress (see Section 1.3) – that is, this value depends on the level of description of the material as a continuous medium. Therefore, the question whether γ_{eff} is a material constant has to be answered experimentally.

To describe the dynamics of fracture, one generalization of this hypothesis is necessary from the very beginning. Obviously, the fracturing process will be different for different velocities of the fracture propagation. Thus, one may expect that γ_{eff} will depend on this velocity. There must exist a range of propagation velocities within which γ_{eff} should decrease with velocity, that is, a range where the fracture process passes from isothermal to adiabatic (more precisely, almost adiabatic) conditions, and γ_{eff} becomes closer to the "true" surface energy γ.

On the other hand, as was indicated in Section 2.1, the process of fracture propagation can be looked upon as a process of nucleation and coalescence of the microfractures that appear in front of the edge of the macrofracture. With an increase in fracture velocity (approaching the velocities of elastic waves) the nucleation and growth of microfractures become more difficult, and this leads to an increase in the "true" surface energy γ. To investigate the behavior of γ_{eff} at lower fracture velocities, the process of nucleation of microfractures must be considered in more detail.

At the lowest (i.e., submolecular) scale level, fracture reduces to the breaking of intermolecular bonds, which may result not only from mechanical work but also from fluctuations in the thermal motion of molecules. In that case, the role of mechanical stress consists only in preventing the restoration of broken bonds. The fluctuational mechanism of bond rupture is a thermally activated process and consequently takes some time, which increases as the temperature and the mechanical stress decrease. Thus, if the fracture is sufficiently slow, a significant part of the surface energy can be supplied by nonmechanical means because of the thermal oscillations. The duration of the fracture process depends on the velocity v of the macrofracture propagation. Thus, at very low velocities, the "true" surface energy γ, in which (as is clear from the foregoing) only mechanical work is included, must decrease, and in principle it must vanish when the fracture velocity approaches zero. Taking into account macroscopic time-dependent processes like stress corrosion, static fatigue, and hydrogen embrittlement will lead to the same conclusion.

This reasoning implies the following qualitative dependence of γ_{eff} on the fracture velocity (schematically represented in Figure 2.5): For extremely slow quasi-static fracture growth, the specific fracture work is close to zero. It increases with an increase in the fracture velocity; then at some velocity γ_{eff} starts decreasing and reaches a minimum. Finally, at velocities comparable to the elastic wave velocities, it once again increases and approaches a limiting value.

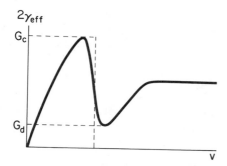

2.5   Fracture energy versus fracture speed.

Let us go back to relation (2.3.13). When γ_{eff} is considered a character-istic function of the fracture propagation velocity (and other thermody-namic characteristics of state, e.g., temperature T) for a given medium, the right side of equation (2.3.13) is not the definition of γ_{eff} since it has another physical meaning: The integral on the right side represents the work done by the medium on the fracture edge to create a unit area of fracture surface. In other words, it is the power per unit length of the edge supplied to move it divided by the propagation velocity. Consider-ing the position of the fracture edge as a generalized coordinate, this integral becomes the generalized force. Following Irwin, let us call it the crack-driving force and denote it by G. The relation (2.3.13) can be rewritten in the form of two equations:

$$G = 2\gamma_{eff} \tag{2.3.14}$$

$$G = -2 \lim_{\varepsilon \to 0} \int_{\xi_2'}^{\xi_2''} \left(\tau_{i1}' u_{i,2}' \right) \Big]_{\xi_1 = -\varepsilon}^{\xi_1 = +\varepsilon} d\xi_2 \tag{2.3.15}$$

The first of these equations states that the crack-driving force is double the effective surface tension, and the second expresses this force in terms of the stress and displacement distributions around the fracture edge.

Relation (2.3.14) is derived for propagating fractures. If the force G is less than $2\gamma_{eff}$, fracture propagation is impossible. However, G cannot exceed $2\gamma_{eff}$ since this is prohibited by the energy conservation law. So the condition (2.3.14) should be rewritten as

$$G \leq 2\gamma_{eff} \tag{2.3.16}$$

where the equality is achieved only for the propagating fracture.

If our hypothetical dependence of γ_{eff} on the propagation velocity is correct (and it is quite plausible since it is based on the most general physical reasoning), the equality in equation (2.3.16) will be satisfied for any value of G, because with decreasing propagation velocity, γ_{eff} tends toward zero. In other words, fracture growth in the medium will take place under any arbitrarily small load. This conclusion, however, requires clarification. Remember that, at the microscopic level, the process of fracture propagation is discrete and consists of elementary events of nucleation and coalescence of microfractures. Here the fracture velocity in question is the average velocity over small but "macroscopic" intervals through which the crack edge passes – that is, intervals several times larger than the size of the elementary particles of the medium. Let us now consider a conventional fracture toughness test. In this case the load on the sample is applied comparatively rapidly, and a fracture has no time to propagate over a "macroscopic" distance. (Remember that a decrease in γ_{eff} occurs due to thermofluctuation effects.) Therefore, in this case, fracture remains at rest until G attains the critical value G_c corresponding to the maximum on the curve $\gamma_{eff}(v)$. At this moment, the fracture starts to propagate dynamically. The reason is that, owing to the decrease in γ_{eff}, the crack-driving force G must also decrease, and this is possible for a fixed size of fracture, due only to an increase in propagation velocity to a value comparable to the velocity of elastic waves. This phenomenon, that is, a sudden (on the laboratory time scale) increase in the fracture velocity, is actually observed during fracture toughness tests on most structural materials and makes G_c a well-defined experimental quantity.

For the case of fracture at an earthquake source, the onset of dynamic propagation, or the onset of the earthquake itself, is associated with the crack-driving force G reaching its maximum value G_c. Now it is seen that the value of G_c, as well as the fracture velocity at which this value is attained, depends on the level of macroscopic description of the medium. In fact, dynamic propagation of "microfractures" in a macroscopic description appears as a slow kinematic growth of a macrofracture. Evidently, the value of G_c for the microlevel is always less than that for a macrolevel description, because the transition to the higher level of "macroscopicity" is related to the incorporation into the fracture work of the losses due to irreversible processes in an even larger volume surrounding the fracture edge. The best known change in G_c (or what is the same, γ_{eff}) occurs when loss due to plastic strain is included in the fracture work, that is, due to the introduction of the Irwin–Orowan concept of "quasi-brittle fracture." According to some estimates, the effective surface energy increases by several orders of magnitude. This level describes well the fracture process for laboratory tests, when the size of representative particles of the medium is a few centimeters.

It is interesting to discuss here the order of magnitude of G_c that is estimated from laboratory experiments in shear as well as from earthquake studies. These estimates are listed in Table 2.1. The seismic estimates of G_c lie between 10^2 and 10^8 joules/m². The scatter in these estimates may reflect, among other things, different physical and chemical conditions on a fault, for example, gouge thickness and pore pressure. The laboratory estimates lie between 0.1 and 10^4 joules/m² and are lower for frictional surfaces than for fracture of intact rock samples and are much larger than similar estimates for tensile cracks in polycrystalline rocks (Atkinson, 1979). The difference between G_c estimated for earthquakes and those measured in polycrystalline rocks in the laboratory is striking. Notwithstanding the scatter in the data, it is clear that the field estimates, in general, are much larger than laboratory estimates. Clearly, a better understanding of both the slip-weakening process in the laboratory and the natural faulting process is needed before these two sets of estimates of G_c can be reconciled. In any case, it should always be kept in mind not only that the laboratory data are representative of a different scale level (different material) than the natural data, but that the latter belong to different scales for different earthquake magnitudes. This could basically account for the above discrepancy and scatter.

As in some engineering applications, the condition for fracture arrest is as important as the condition for the onset of fast propagation. Fracture arrest occurs when under the given conditions (applied stress, fracture size, etc.) the equality in (2.3.16) becomes impossible. Thus, for fracture arrest, either a decrease in G values or a local increase in the strength of the medium γ_{eff} is necessary. In both cases, the decrease in G should be accompanied by a decrease in the fracture velocity, that is, by approaching the minimum on the curve γ_{eff} versus v from the right. When G attains the minimum value G_d (see Figure 2.5), further fracture propagation becomes impossible because the force G_d, determined by external factors, is not sufficient. Thus, three fracture criteria are obtained: a criterion (2.3.16) for propagation, a criterion

$$G = G_c \tag{2.3.17}$$

for the onset of fast propagation, and a criterion

$$G = G_d \tag{2.3.18}$$

for its arrest. Here G_c and G_d are determined, respectively, from the

Table 2.1. *Estimates of* G_c *from earthquake data and laboratory experiments*

G_c (Jm^{-2})	Method of determination and/or experimental conditions	Reference
From earthquake data		
10^7	From fracture propagation time	Takeuchi and Kikuchi (1973)
10^7	From the maximum velocity and accelerations recorded on strong motion instruments	Ida (1973)
10^2	From spatial distribution of slip during fault creep estimated from field data	Rice and Simons (1976)
$1-10^{4a}$ 10^4-10^{6b}	From stress drop and fault dimension	Husseini et al. (1975)
10^6-10^8	From observed fracture velocities for earthquakes combined with theoretical modeling	Das (1976)
10^6-10^8	From stress drop and fault dimension estimated from strong ground motion data, spatial distribution of aftershocks, and geologic evidence	Aki (1979)
10^6	From geodetic data and earthquake recurrence rate for the San Andreas fault	Rudnicki (1980)
10^3	From seismic data from a South African gold mine from McGarr et al. (1979)	Wong (1982)
10^6	From seismic moment and underground measurement of average slip made by McGarr et al. (1979)	Wong (1982)
From laboratory data		
10^3	From postfailure curve at room temperature and 1–800 bars normal stress (on Westerly granite)	Wawersik and Brace (1971)
10^4	At room temperature and 50–450 bars normal stress (on Fichtelgebirge granite)	Rummel et al. (1978)
10^3-10^4	At room temperature and 150–300 bars normal stress (on Witwatersrand quartzite)	Spottiswoode (unpubl.)
10^4	From a slip-weakening model based on the experimental data of Rummel et al. (1978) on saw-cut samples and intact rock (granite) at room temperature and 5 kbar normal stress	Rice (1980)
$0.1-1$	From a slip-weakening model and experiments on saw-cut samples of Sierra white granite at room temperature and ~ 100–400 bars normal stress	Okubo and Dieterich (1981)
10^4	From a slip-weakening model and experiments on intact rock	Wong (1982)

Table 2.1. *(cont.)*

G_c (Jm^{-2})	Method of determination and/or experimental conditions	Reference
10^4	(Westerly granite) at temperatures of 150 to 700°C and 800 bars to 2.5 kbar normal stress	
	From estimates of critical stress intensity factor k_{2c} in pyrophillite specimens with soft inclusions at room temperature and pressure	Sobolev and Rummel (1982)
10	From estimates of k_{3c} from acoustic emission event rates in various rocks using a double torsion method at room temperature and pressure	Cox and Atkinson (1983)
10^2–10^3	From integration of load-displacement record observed at room temperature and pressure	Cox and Scholz (1985)
10^4	From a slip-weakening model and experiments on gabbro at room temperature and ~ 450–600 bars normal stresses	Wong and Biegel (1985)

[a] Frictional.
[b] Fracture.
Source: Modified from Wong (1982).

maximum and minimum values of effective surface energy γ_{eff} as a function of velocity. Conditions (2.3.16) through (2.3.18) express various aspects of Griffith's criterion, the last two following from the condition of fracture propagation (2.3.16) as limiting cases. The arrest criterion was first applied by Husseini et al. (1975) to earthquake faults. Fracture arrest determined from condition (2.3.18) should be understood as an abrupt drop in the speed of propagation from dynamic to quasi-static values. This means possible further slow growth, which again has a tendency to become discrete and discontinuous on the normal time scale (aftershocks).

Crack-driving force for an isotropic, elastic, ideally brittle medium
   The stress and strain distribution in the vicinity of the fracture edge depends only on the rheological properties of the medium, mode of fracture, and fracture velocity. It does not depend, up to constant factors, on the size of the fracture, configuration of the body, or

distribution of external loads. Therefore, for a given rheology, it is possible to obtain a general solution for this distribution and, on this basis, to specify the expression (2.3.15) for the crack-driving force. Only the results for isotropic elastic medium are presented here. Because of the linear relation between stress and strain, τ_{ij} and $u_{i,j}$ must have singularities of the same order at the fracture edge. If the crack-driving force does not vanish, the integrand in (2.3.15), as indicated above, should have a singularity of the order 1, and consequently τ_{ij} and $u_{i,j}$ should have singularities of the order $\frac{1}{2}$. In particular, the stress components ahead of the fracture edge should have the form

$$\tau'_{i1} = \frac{k_i}{\sqrt{2\xi_2}} \qquad \text{for} \quad \xi_2 > 0 \qquad (2.3.19)$$

It can be shown that equation (2.3.12), together with Hooke's law, permits a solution with such a singularity only under the condition that the fracture velocity does not exceed the velocity of shear waves, v_S, and hence, of course, the velocity of longitudinal waves v_P as well, where

$$v_S = \left(\frac{\mu}{\rho}\right)^{1/2}; \qquad v_P = \frac{(\lambda + 2\mu)^{1/2}}{\rho} \qquad (2.3.20)$$

and that the frictional stress between the fracture faces is finite near its edge. In particular, for viscous friction, the order of the stress singularity is less than $\frac{1}{2}$ and the force G vanishes for any nonvanishing fracture velocity. This means that fracture propagation in an ideally brittle medium under viscous friction is impossible; that is, the model of an ideally brittle body is incompatible with viscous friction. If the frictional traction is finite, then the solution with the necessary singularity can be obtained and expression (2.3.15) takes the form (Kostrov, Nikitin, and Flitman, 1969)

$$G(v, k_1, k_2, k_3) = \frac{\pi v}{2\mu\Delta(v)v_S^2}\left[\left(1 - \frac{v^2}{v_P^2}\right)^{1/2}k_1^2 + \left(1 - \frac{v^2}{v_S^2}\right)^{1/2}k_2^2\right]$$

$$+ \frac{\pi}{2\mu\left(1 - v^2/v_S^2\right)^{1/2}}k_3^2 \qquad (2.3.21)$$

for $v < v_S$ and $G = 0$ otherwise. Here $\Delta(v)$ is the left side of the

Rayleigh equation $\Delta(v) = 0$ for the Rayleigh wave velocity v_R:

$$\Delta(v) = 4\left(1 - \frac{v^2}{v_P^2}\right)^{1/2}\left(1 - \frac{v^2}{v_S^2}\right)^{1/2} - \left(2 - \frac{v^2}{v_S^2}\right)^2 \tag{2.3.22}$$

For shear fractures, $k_1 = 0$ in expression (2.3.21).

Analysis of expression (2.3.21) leads to an important conclusion on the possible values of fracture propagation velocity in an ideally brittle elastic medium. According to the definition of γ_{eff} as the total specific fracture work, this value has to be positive. Therefore, according to expression (2.3.14) the necessary condition for fracture propagation is the positiveness of G,

$$G(v, k_1, k_2, k_3) > 0 \tag{2.3.23}$$

Consider the expression (2.3.21) from this point of view. The k_3 term is positive and increases infinitely as $v \to v_S$. The sign of the k_1 and k_2 terms is determined from the sign of the function $\Delta(v)$. As $v \to 0$, the value of Δ also tends toward zero, and the leading term of the power series expansion in v has the form

$$\Delta = 2v^2\left(\frac{1}{v_S^2} - \frac{1}{v_P^2}\right) \qquad \text{for} \quad v \to 0 \tag{2.3.24}$$

Thus, for small v the value of Δ, and correspondingly the whole expression (2.3.21), is positive.

Rewriting the leading term of this expression for small v (i.e., $v \ll v_S$), which is obtained by substituting expression (2.3.24) into (2.3.21) and replacing the radicals in the latter by 1, we obtain

$$G = \frac{\pi(1 - \nu^2)}{E}(k_1^2 + k_2^2) + \frac{\pi}{2\mu}k_3^2 \qquad \text{for} \quad v \ll v_S \tag{2.3.25}$$

where v_P and v_S are expressed in terms of Young's modulus E and Poisson's ratio ν. This is the well-known Irwin formula. It is this expression that must be used in the criteria for the onset of fast propagation and arrest, (2.3.17) and (2.3.18). For every "pure" fracture mode, tensile ($k_1 \neq 0$, $k_2 = k_3 = 0$), inplane shear ($k_2 \neq 0$, $k_1 = k_3 = 0$), and antiplane shear ($k_3 \neq 0$, $k_1 = k_2 = 0$), using (2.3.17), (2.3.18), and (2.3.25), Irwin's fracture onset criteria are obtained as

$$k_i \le k_{ic} \tag{2.3.26}$$

where

$$\frac{\pi(1 - \nu^2)}{E} k_{1c}^2 = G_c(k_{1c}, 0, 0) \equiv G_{1c}$$

$$\frac{\pi(1 - \nu^2)}{E} k_{2c}^2 = G_c(0, k_{2c}, 0) \equiv G_{2c} \qquad (2.3.27)$$

$$\frac{\pi}{2\mu} k_{3c}^2 = G_c(0, 0, k_{3c}) \equiv G_{3c}$$

The corresponding criteria for fracture arrest have a similar form. Here, as well as in (2.3.21), the arguments in $G_c(k_1, k_2, k_3)$ emphasize the dependence of G_c on the fracture mode.

As was stated in Section 1.3, it is physically quite possible that one and the same surface may be fractured in shear and not in tension. This suggests that the three modes of fracture are not necessarily physically dependent. Consequently, one must, strictly speaking, consider not a single crack edge but three crack edges, which possibly but not necessarily coincide. Accordingly, it is convenient to split the crack-driving force G into three terms, $G = G_1 + G_2 + G_3$ corresponding to the tensile, inplane, and antiplane modes of fracture, respectively. From (2.3.21), we obtain

$$G_1 = \frac{\pi v \left(1 - v^2/v_P^2\right)^{1/2}}{2\mu\Delta(v) v_S^2} k_1^2$$

$$G_2 = \frac{\pi v \left(1 - v^2/v_S^2\right)^{1/2}}{2\mu\Delta(v) v_S^2} k_2^2 \qquad (2.3.28)$$

$$G_3 = \frac{\pi}{2\mu \left(1 - v^2/v_S^2\right)^{1/2}} k_3^2$$

Because energy is consumed in fracture and there is no physical reason for arguing against a crack surface being fractured in a given mode and not in the other two modes, an important conclusion is that each of these terms must be nonnegative. This will be true even when the surface is fractured in more than one mode. The function $\Delta(v)$ is negative when $v \leq v_R$, so G_1 and G_2 for such crack velocities must vanish, which is possible only if $k_1 = k_2 = 0$. Consequently, if a crack propagates with super-Rayleigh velocity, the fracture at its edge may occur only in Mode III. If the crack conserves some toughness (strength) against Mode I and/or Mode II fracture, these will occur separately and propagate at sub-Rayleigh velocities.

All the conclusions on possible crack propagation velocities reached here depend on the validity of the ideally brittle model, that is, when the process zone size can be neglected. In nonideally brittle bodies, shear cracks may propagate with any velocity up to the speed of longitudinal waves. This was shown analytically by Burridge, Conn, and Freund (1979) and numerically by Andrews (1976a, 1985), Das and Aki (1977), Das (1981), Day (1982a), and Okubo (1986). Laboratory experiments on rocks (Johnson and Scholz, 1976) also sometimes show transonic shear wave crack speeds. This is understandable, since the process zone cannot be smaller than the grain size, and for shear fracture one may expect it to be much larger.

2.4 Conclusions to Part I

In Part I, we have discussed the mechanics of the continuous medium and fracture mechanics as applied to earthquake sources. The equations necessary to formulate the complete problem of spontaneous fracture propagation have been developed for a perfectly brittle model of fracture.

One could hardly expect to solve this problem by analytic methods. Even if the fracture criterion were discarded and the fracture edge propagation history were preassigned, the existing methods would enable one to solve the resulting problem of dynamic theory of elasticity only for some of the simplest cases (e.g., plane fracture surface Σ, simple shaped cracks – penny-shaped or elliptical – constant fracture velocity, and uniform initial stress). Therefore, progress will be made in this field only through the application of direct numerical methods, which is, however, a difficult task and will be discussed in Chapter 6. These difficulties show that it is just as important to develop some qualitative conclusions from the problem formulations and fundamental physical assumptions as it is to solve the problem itself.

The most fundamental result here is the conclusion regarding the maximum fracture velocity (see Section 2.3). A no less important conclusion in fracture mechanics concerns the relation between the "stress drop" and the initial fracture size, which is discussed in the following paragraphs.

It seems reasonable to assume, at least for medium-sized earthquakes, that the cause of tectonic stress, whatever we imagine it, is situated far from the earthquake source compared with the size of the initial fracture.

Then the initial tectonic stress can be taken as uniform far from the initial fracture. The size of the final fracture after the earthquake must be many times larger than that of the initial defect. Consequently, the average initial stress on the final fracture must coincide with this uniform stress σ^∞. If the interaction of the fault faces is described by the dry friction law, the resulting traction on the fault must be equal to the kinetic friction since the initial fault was slipping, even if slowly. Since we use the linear elastic theory, subtracing the homogeneous stress from the actual stress, we obtain a crack problem for the frictionless crack and the applied (at infinity) stress equal to the difference between initial state and friction $(\sigma^\infty - g)$, where g is frictional stress (principle of super-position). The dimension of the stress intensity factor is (stress) $\times$ (length)$^{1/2}$. Since the stress intensity factors must be proportional to $(\sigma^\infty - g)$, we obtain, taking the size L_0 of the initial defect as the scaling length,

$$k_i = (\sigma^\infty - g)\sqrt{L_0}\,\lambda_i \qquad\qquad (2.4.1)$$

where the λ_i are dimensionless functions of the position of the fracture edge. Substituting this expression into (2.3.25), we get

$$G_c = \frac{\pi(\sigma^\infty - g)^2 L_0}{\mu}\lambda_G \qquad\qquad (2.4.2)$$

where λ_G is a dimensionless function. The difference $(\sigma^\infty - g)$ is equal to the average stress drop $\Delta\sigma$ on the fracture surface.

Thus, from (2.4.2), omitting the factor λ_G we get

$$\Delta\sigma = \left(\frac{\mu G_c}{\pi L_0}\right)^{1/2} \qquad\qquad (2.4.3)$$

$$\Delta\sigma L_0^{1/2} = \text{const} \qquad\qquad (2.4.4)$$

where the constant on the right side depends on the rigidity μ and toughness G_c of the medium. This discussion is a simple extension of the explanation of strength value in fracture mechanics to the specific case of earthquake source mechanics.

Next, if we assume that the value of G_c for rock samples and earthquakes is roughly the same (i.e., if we discard the difference in the level of description of the medium – the scale factor), then we may come to some conclusions on the relation of $\Delta\sigma$ to the strength value σ_B, for laboratory samples. This same reasoning leads to the expression of the same form as (2.4.4):

$$\sigma_B l^{1/2} = \text{const} \tag{2.4.5}$$

where $l^{1/2}$ is the size of the initial defect in the sample of the order of grain size, and the constant on the right side is, as assumed, of the same order of magnitude as the constant in (2.4.4). Thus,

$$\Delta\sigma/\sigma_B = (l/L_0)^{1/2} \tag{2.4.6}$$

The grain size l in rocks is of the order of millimeters. The fracture size at an earthquake source varies from hundreds of meters to tens and hundreds of kilometers, that is, 10^5 to 10^7 times larger than l. Therefore, from (2.4.6) we get

$$\log(\Delta\sigma/\sigma_B) = 2.5 \pm 1 \tag{2.4.7}$$

depending on the earthquake magnitude. The shear strength is estimated as 10^9dyn/cm^2 for rocks. Therefore, the stress drop at the source of a sufficiently large earthquake can be estimated as

$$\Delta\sigma = 3 \times 10^6 \ \text{dyn/cm}^2 = 3 \ \text{bars} \tag{2.4.8}$$

This agrees with the stress drop estimates obtained for some earthquakes, but this agreement should not be overemphasized because it is based on very rough assumptions.

The initial defects for a given earthquake are created by earlier, smaller earthquakes. Therefore, if we assume that the size L_0 of the initial fracture is larger for larger earthquakes than for smaller ones, it follows from (2.4.4) that the stress drop at the earthquake source should be higher for small and lower for large earthquakes, *if all the other conditions are the same*. This conclusion requires further discussion. It is assumed that the size of an earthquake is determined by the size of the initial defect, with friction and G_c being the same for small and large events. But "initial defects" are the result of smaller earthquakes and are finite in size only because of their arrest by some strength barriers (regions of increased G_c), so to reinitiate them it would be necessary to increase the applied stress σ^∞. This consideration would lead to the opposite conclusion, namely, that the larger the earthquake the greater is the stress drop! Empirically, it was found (first by Aki, 1967, and later

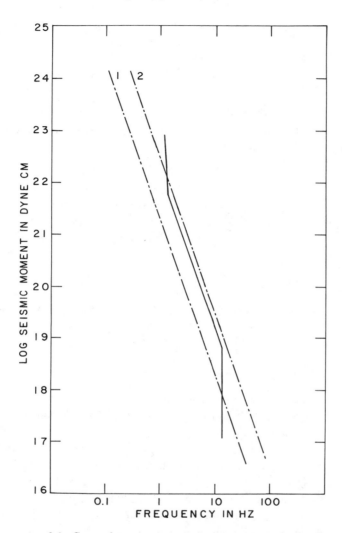

2.6   Corner frequency versus seismic moment for earthquakes in California.
Key: 1, Aki (1967, 1972b): extrapolated from large earthquakes. 2, Hanks
(1976): extrapolated from large earthquakes in southern California. (*Adapted
from Chouet et al.*, 1978. © *Seismol Soc. Am.*)

confirmed by others) that the stress drop remains essentially constant for
a large range of magnitudes, which suggests that larger earthquakes are
initiated at even higher strength barriers that arrest rupture propagation
during preceding smaller earthquakes.

As mentioned earlier, equation (2.4.5) when applied to polycrystalline materials gives l of the order of the grain size. This is understandable since the grain boundaries represent the strongest strength barriers in a sample. It seems that the deviations from Aki's scaling law (Aki, 1972b; Chouet, Aki, and Tsujiura, 1978) observed for magnitudes of ≈ 5 and ≈ 2 or 3 (Figure 2.6) have a similar nature, revealing natural scales of the earth's crustal structure. Namely, at these magnitudes the size of the source (reflected in the corner frequency value) does not increase with magnitude. In fact, this is the only definitive seismological evidence on the existence of the hierarchical structure of the earth material discussed at length in this part of the book.

The occurrence of an earthquake is, even now, sometimes related to the accumulated tectonic stress reaching the strength value for rocks. As seen in this chapter, the strength of a sample of a material is determined not only fracture work γ_{eff} (or G_c), but also by the sizes of initial cracks within the sample. That the strength measurements in laboratory tests give constant stress value for a given material is explained by the selection of samples when samples with appreciable defects are rejected. In engineering, the possibility of using (laboratory) strength values thus obtained is due to the use of materials in which one cannot detect defects. For earthquakes, however, we have to deal with rocks that have not been subjected to such selection and, as seen from geologic and seismological data, may contain arbitrarily large discontinuities of up to several tens or hundreds of kilometers. In these circumstances, the concept of strength as a material property becomes meaningless. Therefore, laboratory tests of strength are useless for investigations of earthquake sources. Moreover, because as the condition for the onset of fast propagation $G = G_c$ involves not only the value of stress but also the fracture size, earthquakes might be initiated not only by stress accumulation but also by an increase in the length L_0 of initial fracture by slow fracture growth. It is seen that the accumulation of tectonic stress is, generally speaking, not necessary for the initiation of an earthquake. What is necessary is a certain stress level to maintain slow (kinetic) fracture growth. The latter, as has been mentioned, tends to be discrete and manifests as background seismicity.

These discussions show that perhaps the most promising approach to earthquake prediction might be based on the variations of background seismicity preceding large earthquakes. This subject, however, is beyond the framework of this book.

Part II

INVERSION FOR SOURCE PARAMETERS

3

THE INVERSE PROBLEM OF EARTHQUAKE SOURCE THEORY

Information on the earthquake faulting process has to be obtained from the analysis of seismic waves radiated from the source. To extract this information, it is necessary to establish how the seismic radiation is related to the characteristics of the source. The concepts of fracture mechanics as presented in Part I provide some understanding of the physical meaning of the fracture at the source and permit certain general conclusions to be drawn about the specific features of the process. However, the dynamic description of fracture, based on fracture mechanics, leads to the boundary value problems of the dynamic theory of elasticity, which are unsolvable in general form. Consequently, with this description the relation of seismic radiation to the process of fracture propagation cannot be investigated with the generality necessary for formulating the inverse problem, that is, that of obtaining information about the source from seismic observations. The kinematic description in terms of the displacement jump vector on the fracture surface as a function of position and time is more advantageous from this point of view because, in this case, the most general solution to the problem of radiation exists, permitting the inverse problem formulation.

In the foregoing remarks, it is implicitly assumed that the part of the fault where nonvanishing slip occurs during an earthquake has finite dimensions. Physically this is always true, for all existing faults have finite dimensions. A situation is possible, however, in which the major part of the fault creeps quasi-statically, and the stress is accumulated and then released during earthquakes only on some locked areas of the fault. If an earthquake occurs due to the breaking of a single such patch ("asperity") whose dimension is much less than that of the whole fault, then when considering the seismic radiation at distances that are small compared with the fault size but large compared with the asperity size, one can consider the fault to be infinite. Thus, one obtains the "asperity model" for an earthquake source, which has recently become popular. If the problem for the asperity model were formulated in terms of the displacement jump, the very concept of the earthquake source would fail since its dimensions would be infinite. On the other hand, the stress drop distribution on the fault is confined to the area of the asperity, and by formulating the problem in terms of stress drop one can again obtain a source of seismic radiation having finite size. It should be pointed out that such an asperity model will provide an adequate description of the seismic radiation even at large distances compared with the size of the whole fault but in the vicinity of first arrivals, when the diffracted waves from the fault edge have not yet arrived.

This chapter is devoted to the classical "dislocation" or "crack" model of the earthquake source. In Section 4.6 the seismic radiation due to the asperity model is considered in some detail.

3.1   Green–Volterra formula; Green's tensor in linear elastodynamics

For a linearly elastic medium occupying volume V not containing discontinuities, consider two solutions: $u_i(\mathbf{x}, t)$ and $u_i'(\mathbf{x}, t)$ of the equations of elasticity corresponding to two different distributions of body forces f_i and f_i'. We start with the equations of motion and Hooke's law for these solutions.

$$\sigma_{ij,j} + \rho f_i = \rho \ddot{u}_i \tag{3.1.1}$$

$$\sigma_{ij} = c_{ijkl} u_{k,l} \tag{3.1.2}$$

$$\sigma_{ij,j}' + \rho f_i' = \rho \ddot{u}_i' \tag{3.1.1'}$$

$$\sigma_{ij}' = c_{ijkl} u_{k,l}' \tag{3.1.2'}$$

Multiplying (3.1.1) by u_i' and (3.1.1') by u_i and subtracting, we get

$$\sigma_{ij,j} u_i' - \sigma_{ij,j}' u_i + \rho(f_i u_i' - f_i' u_i) - \rho(\ddot{u}_i u_i' - \ddot{u}_i' u_i) = 0$$

Using (3.1.2) and (3.1.2'), we can rewrite this equation as

$$(\sigma_{ij} u_i' - \sigma_{ij}' u_i)_{,j} - \frac{\partial}{\partial t} \rho(\dot{u}_i u_i' - \dot{u}_i' u_i) + \rho(f_i u_i' - f_i' u_i) = 0 \tag{3.1.3}$$

Here, the symmetry of the stiffness tensor is used:

$$c_{ijkl} = c_{klij}$$

We integrate (3.1.3) over a four-dimensional region Ω, which is the direct product of the volume V and the time interval $0 \le t \le t_1$. The first two terms in equation (3.1.3) are the four-dimensional divergence, and hence the integral of these can be transformed into the integral over the boundary Γ of the region Ω. Thus, we get

$$\int_\Gamma \left\{ (\sigma_{ij} u_i' - \sigma_{ij}' u_i) N_j - \rho(u_i u_i' - u_i' u_i) N_{(t)} \right\} d\Gamma$$

$$= \int_\Omega \rho(f_i' u_i - f_i u_i') \, d\Omega \tag{3.1.4}$$

where $[N_j, N_{(t)}]$ is the four-dimensional external normal to Ω.

If the volume V is independent of time, the integrals over Ω and Γ can be rewritten as double integrals. Then the boundary Γ is split into three parts: (1) the volume V at time $t = 0$, (2) the volume V at time $t = t_1$, and (3) the direct product of the boundary S of the volume V and time interval $0 \le t \le t_1$. On the first and second parts of Γ, the components of the four-dimensional normal would be $N_j = 0$, $N_{(t)} = -1$ (for $t = 0$), or $N_{(t)} = 1$ (for $t = t_1$). On the third part of the boundary Γ, the components of the normal would be $N_j = n_j$, $N_{(t)} = 0$, where n_j is the three-dimensional external normal to S. Thus, equation (3.1.4) can be rewritten as

$$\int_0^{t_1} dt \int_S (\sigma_{ij} u_i' - \sigma_{ij}' u_i) n_j \, dS + \int_0^{t_1} dt \int_V \rho(f_i u_i' - f_i' u_i) \, dV$$

$$+ \int_V \rho(u_i \dot{u}_i' - \dot{u}_i u_i') \, dV \Big|_{t=0}^{t=t_1} = 0 \tag{3.1.5}$$

Equation (3.1.5) is called the Green–Volterra formula and is a generalization of Betti's reciprocity theorem for elastodynamics. It is important to note that the relation (3.1.5) is also applicable to the case where one of the solutions u_i or u_i' is a distribution, provided that the other is sufficiently smooth.

Consider the solutions U_{ik} corresponding to three concentrated unit forces directed along the coordinate axes:

$$\rho f_i' = \delta_{ik} \delta(t - t_0) \delta(\mathbf{x} - \mathbf{x}_0), \qquad k = 1, 2, 3 \tag{3.1.6}$$

and satisfying the initial conditions

$$U_{ik} = \dot{U}_{ik} = 0 \qquad \text{for} \quad t \le t_0 \tag{3.1.7}$$

and the boundary conditions

$$\Sigma_{ikj} n_j = 0 \qquad \text{for} \quad \mathbf{x} \in S, \qquad t \geq t_0 \tag{3.1.8}$$

where δ_{ik} is the Kronecker symbol, $\delta(t)$ is the one-dimensional and $\delta(\mathbf{x})$ the three-dimensional Dirac delta function, and Σ_{ikj} are the stress tensors corresponding to U_{ik}.

Since the equations of motion are homogeneous with respect to time, these solutions depend only on the difference $t - t_0$, but not on t and t_0 separately. Let us denote these three solutions by $U_{ik}(\mathbf{x}, \mathbf{x}_0, t - t_0)$, and the corresponding stress tensors by $\Sigma_{ijk}(\mathbf{x}, \mathbf{x}_0, t - t_0)$. Here, the last index in Σ and U indicates that the solution corresponds to a concentrated force directed along the axis x_k. From Hooke's law, we get

$$\Sigma_{ijk} = c_{ijlm} U_{lk, m} \tag{3.1.9}$$

The three vectors U_{ik} ($k = 1, 2, 3$) together represent a tensor of second order with respect to the subscripts i and k. This tensor is called Green's tensor for elastic problems with stress-type boundary conditions.

A detailed study of the properties of Green's tensor in an inhomogeneous, anisotropic medium is not within the scope of this book. Let us note only that $U_{ik}(\mathbf{x}, \mathbf{x}_0, t - t_0)$ is a smooth function of its arguments everywhere outside of some surfaces – fronts of elastic waves produced by concentrated forces (3.1.6).

Equations (3.1.1') and (3.1.2') are invariant with respect to the change of the sign of time. Therefore, the tensor $U_{ik}(\mathbf{x}, \mathbf{x}_0, t_0 - t)$ is also the solution for the body forces (3.1.6) and satisfies the boundary conditions (3.1.8). However, instead of the initial conditions (3.1.7), we have in this case

$$\dot{U}_{ik}(\mathbf{x}_1, \mathbf{x}_0, t_0 - t) = U_{ik}(\mathbf{x}, \mathbf{x}_0, t_0 - t) = 0 \qquad \text{for} \quad t > t_0 \tag{3.1.10}$$

Apply the relation (3.1.5) to the two solutions $u_i = U_{ik}(\mathbf{x}, \mathbf{x}_0, t)$ and $u_i' U_{il}(\mathbf{x}, \mathbf{x}_1, t_1 - t)$. In view of the boundary conditions (3.1.8) the first integral in (3.1.5) will vanish. Similarly, due to the initial conditions (3.1.7) and (3.1.10), the third term in (3.1.5) will also vanish. As a result, we get

$$\int_0^{t_1} dt \int_V \{ \delta_{ik} \, \delta(t) \, \delta(\mathbf{x} - \mathbf{x}_0) U_{il}(\mathbf{x}, \mathbf{x}_1, t_1 - t)$$
$$- \delta_{il} \, \delta(t_1 - t) \, \delta(\mathbf{x} - \mathbf{x}_1) U_{ik}(\mathbf{x}, \mathbf{x}_0, t) \} \, dV = 0$$

Here the integrals are eliminated due to δ functions, and we obtain

$$U_{kl}(\mathbf{x}_0, \mathbf{x}_1, t_1) = U_{lk}(\mathbf{x}_1, \mathbf{x}_0, t_1) \tag{3.1.11}$$

This is the well-known reciprocity theorem of elastodynamics.

The solution for Green's tensor for a given volume V and a given distribution of density ρ and stiffness coefficients c_{ijkl} is, in general, a very difficult task. However, if this problem is already solved, the solution of any other particular problem can easily be expressed through Green's tensor. Consider, for example, the solution u_i corresponding to $f_i = 0$ and satisfying the homogeneous boundary conditions $\sigma_{ij}n_j = 0$ at S but inhomogeneous initial conditions $u_i = u_i^0(x)$, $\dot{u}_i = v_i^0(x)$ at $t = 0$. Then applying the Green–Volterra formula to the required solution u_i and to Green's tensor $u_i' = U_{ik}(x, x_0, t_1 - t)$ and evaluating the integral with δ functions, we get

$$u_k(x_0, t_1) = \int_V \left\{ u_i^0(x) \frac{\partial}{\partial t_1} U_{ik}(x, x_0, t_1) + v_i^0 U_{ik}(x, x_0, t_1) \right\} dV$$

(3.1.12)

Similarly, the solution for the given conditions at the boundary S or for body force f_i can also be obtained. It will be shown in the next section that the general solution of the fracture problem with kinematic description (dislocation problem) can be as easily obtained using Green's tensor.

3.2   General solution to the kinematic dislocation problem; formulation of the inverse problem

For the sake of convenience, let us reproduce the equations determining the elastic field due to a discontinuity at the earthquake source, for the kinematic description of fracture. Stress perturbation τ_{ik} satisfies the equation of motion (1.3.20),

$$\tau_{ik,k} = \rho \ddot{u}_i$$

(3.2.1)

and is related to the displacement by Hooke's law (1.3.21):

$$\tau_{ik} = c_{iklm} u_{l,m}$$

(3.2.2)

Boundary conditions on the earth's surface (1.3.24) are

$$\tau_{ik} n_k = 0$$

(3.2.3)

Displacement vector u_i satisfies the homogeneous initial conditions (1.3.23),

$$u_i = \dot{u}_i = 0 \qquad \text{for} \quad t = 0$$

(3.2.4)

where $t = 0$ is the time of the earthquake onset.

The condition of traction continuity must be satisfied on the fracture surface $\Sigma(t)$ and the displacement jump $a_i(x, t)$, that is, condition

(1.3.25), must be given:

$$\tau_{ik}^{+} n_k - \tau_{ik}^{-} n_k = 0$$

$$u_i^{+} - u_i^{-} = a_i(\mathbf{x}, t)$$

(3.2.5)

where n_k is the unit normal to $\Sigma(t)$.

In this section we will assume that $U_{ik}(\mathbf{x}, \mathbf{x}_0, t - t_0)$ is Green's tensor for the whole earth in the absence of fractures, which satisfies the homogeneous conditions (3.1.8) at the earth's surface. Formula (3.1.5) cannot be used directly for solving the problem (3.2.1) through (3.2.5) because in this case the displacement u_i is discontinuous on the fracture surface $\Sigma(t)$. To overcome this difficulty, let us extend the fracture surface $\Sigma(t)$ to obtain a piecewise smooth stationary surface Σ, which intersects the earth's surface so that the entire volume V of the earth is divided into two parts V^+ and V^- located on opposite sides of Σ. Across the extension of $\Sigma(t)$, the traction and displacement vector must obviously be continuous; that is, they must satisfy conditions (3.2.5) with $a_i \equiv 0$. Now neither of these volumes V^+ and V^- contains a discontinuity in displacement, and formula (3.1.5) can be applied to them with $u_i' = U_{ik}(\mathbf{x}, \mathbf{x}_1, t_1 - t)$. In view of initial conditions (3.2.4) and (3.1.10), the last term vanishes. Similarly, the integrals over the parts of the earth's surface enclosing V^+ and V^-, respectively, would vanish due to boundary conditions (3.2.3) and the corresponding conditions for Green's tensor.

As a result, for $V^{\pm}$ we get the formula

$$\varepsilon^{\pm} u_k(\mathbf{x}_1, t_1) = \pm \int_0^{t_1} dt \int_{\Sigma^{\pm}} \left[\Sigma_{ijk}(\mathbf{x}, \mathbf{x}_1, t_1 - t) u_i(\mathbf{x}, t) \right.$$

$$\left. - \tau_{ij}(\mathbf{x}, t) U_{ik}(\mathbf{x}, \mathbf{x}_1, t_1 - t) \right] n_j \, dS$$

(3.2.6)

where $\varepsilon^+ = 1$ for $\mathbf{x}_1$ in V^+ and $\varepsilon^+ = 0$ for $\mathbf{x}_1$ in V^-, but $\varepsilon^- = 1$ for $\mathbf{x}_1$ in V^- and $\varepsilon^- = 0$ for $\mathbf{x}_1$ in V^+. Here the fact that on Σ^+ the external normal to V^+ differs in sign from the positive normal n_j to Σ whereas at Σ^- the external normal to V^- coincides with n_j has been taken into account. Green's tensor U_{ik} and the corresponding stress tensor Σ_{ijk} are continuous across Σ; that is, their limiting values on Σ^+ and Σ^- sides of Σ coincide (under the condition that $\mathbf{x}_1$ does not belong to Σ). This is also true for $\tau_{ij} n_j$ in view of the first condition of (3.2.5). Therefore, the relation (3.2.6) can be rewritten in the form

$$\varepsilon^{\pm} u_k(\mathbf{x}_1, t_1) = \pm \int_0^{t_1} dt \int_{\Sigma} \left[\Sigma_{ijk}(\mathbf{x}, \mathbf{x}_1, t_1 - t) u_i^{\pm}(\mathbf{x}, t) \right.$$

$$\left. - \tau_{ij}(\mathbf{x}, t) U_{ik}(\mathbf{x}, \mathbf{x}_1, t_1 - t) \right] n_j \, dS \qquad (3.2.7)$$

Now the sum of these expressions is

$$u_k(\mathbf{x}_1, t_1) = \int_0^{t_1} dt \int_{\Sigma} \Sigma_{ijk}(\mathbf{x}, \mathbf{x}_1, t_1 - t) \left[u_i^{+}(\mathbf{x}, t) - u_i^{-}(\mathbf{x}, t) \right] n_j \, dS$$

$$(3.2.8)$$

Thus, the displacement vector at any point not belonging to Σ is expressed in terms of the displacement jump on Σ. Taking into account condition (3.2.5) and the fact that on the extension of $\Sigma(t)$ the displacement vector is continuous, we finally get

$$u_k(\mathbf{x}_1, t_1) = \int_0^{t_1} dt \int_{\Sigma(t)} \Sigma_{ijk}(\mathbf{x}, \mathbf{x}_1, t_1 - t) a_i(\mathbf{x}, t) n_j(\mathbf{x}) \, dS \qquad (3.2.9)$$

This expression is the general solution of problems (3.2.1) through (3.2.5), in terms of Green's tensor U_{ik} for the whole earth. If we substitute expression (3.1.9) for Σ_{ijk} in (3.2.9), we obtain

$$u_k(\mathbf{x}_1, t_1) = \int_0^{t} dt \int_{\Sigma(t)} \left[\frac{\partial}{\partial x_m} U_{lk}(\mathbf{x}, \mathbf{x}_1, t_1 - t) \right] c_{ijlm}(\mathbf{x}) n_j(\mathbf{x}) a_i(\mathbf{x}, t) \, dS$$

$$(3.2.10)$$

Using the reciprocity theorem (3.1.11), we can rewrite this expression as

$$u_k(\mathbf{x}_1, t_1) = \int_0^{t_1} dt \int_{\Sigma(t)} \left[\frac{\partial}{\partial x_m} U_{kl}(\mathbf{x}_1, \mathbf{x}, t_1 - t) \right] c_{ijlm}(\mathbf{x}) n_j(\mathbf{x}) a_i(\mathbf{x}, t) \, dS$$

$$(3.2.11)$$

This shows that the elastic field $u_k(\mathbf{x}_1, t)$ due to dislocation, as observed at point $\mathbf{x}_1$ and time t_1, coincides with the field that would be produced by point sources distributed on the surface $\Sigma(t)$ in the absence of discontinuities. Thus, the contribution of each element dS of the dislocation surface during time dt to the total field is equal to

$$du_k(\mathbf{x}_1, t) = \left[\frac{\partial}{\partial x_m} U_{kl}(\mathbf{x}_1, \mathbf{x}, t_1 - t) \right] c_{ijlm}(\mathbf{x}) n_j(\mathbf{x}) a_i(\mathbf{x}, t)\, dt\, dS$$

$$(3.2.12)$$

That is, each element is equivalent to a point source. To clarify the type of this point source, we note that $U_{kl}(\mathbf{x}_1, \mathbf{x}, t_1 - t)$ is the displacement produced at time t_1 at point $\mathbf{x}_1$ by the concentrated unit force, applied at instant t at point $\mathbf{x}$ and directed along the axis x_l. Then the derivative

$$\frac{\partial}{\partial x_m} U_{kl}(\mathbf{x}_1, \mathbf{x}, t_1 - t)$$

is the displacement at $(\mathbf{x}_1, t_1)$ due to the unit couple that is applied at $(\mathbf{x}, t)$, where the forces are directed along the axis x_l, and the arm of the couple along the axis x_m. Consequently, the entire expression (3.2.12) represents the radiation of the nine couples that correspond to all combinations of the subscripts of l and m, where the contribution of each of the couples is given by

$$m_{lm}(\mathbf{x}, t) = c_{ijlm}(\mathbf{x}) n_j(\mathbf{x}) a_i(\mathbf{x}, t) \qquad (3.2.13)$$

The nine concentrated couples together are called the elastic dipole. The matrix giving the contribution of each couple is called the dipole moment tensor. Thus, the field due to each element of dislocation is equivalent to the effect of a concentrated dipole having moment tensor $dM_{lm} = m_{lm}\, dS\, dt$. The field of the whole dislocation is equivalent to the field of dipoles, distributed over the surface $\Sigma(t)$ having moment tensor density m_{lm} and can be found from the relation (3.2.13) as

$$u_k(\mathbf{x}_1, t_1) = \int_0^{t_1} dt \int_{\Sigma(t)} m_{lm}(\mathbf{x}, t) \frac{\partial}{\partial x_m} U_{kl}(\mathbf{x}_1, \mathbf{x}, t_1 - t)\, dS \qquad (3.2.14)$$

The dipole moment tensor should not be confused with the total mechanical rotational moment of the couples comprising the dipole. The mechanical moment of each couple is equal to the vector product of force by the arm. Consequently, the total mechanical moment corresponding to the dislocation element is

$$dM_k = e_{klm} m_{lm}\, dS\, dt \qquad (3.2.15)$$

where e_{klm} is the unit antisymmetric tensor. It can be seen from expression (3.2.13) that the tensor m_{lm} is symmetric due to the symmetry

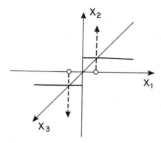

3.1   Dipole representation of a shear dislocation.

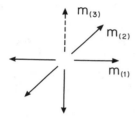

3.2   Dipole representation of a shear dislocation in principal axes.

of the stiffness tensor in l and m. Then (3.2.15) vanishes:

$$dM_k = 0 \qquad\qquad (3.2.16)$$

Thus, the radiation due to a dislocation is equivalent to the radiation due to elastic dipoles without mechanical moment, distributed over the surface $\Sigma(t)$ with a density given by (3.2.13). The body-force equivalents for the kinematic description of the seismic source were first obtained by Burridge and Knopoff (1964).

To illustrate this, consider the couples acting on the plane (x_1, x_2). These couples correspond to the components m_{12} and m_{21}. The couple corresponding to the component m_{12} is represented in Figure 3.1 by solid arrows, and the couple corresponding to the component m_{21} is shown by dashed arrows. It is seen that they balance one another.

Because the tensor m_{lm} is symmetric, it can always be transformed to principal axes. Taking the coordinate axes along the principal ones, we get $m_{lm} = m_{(l)}\delta_{lm}$, where $m_{(l)}$ is the principal value. Each diagonal component corresponds to a couple without moment, which is directed along the respective axis (Figure 3.2). Therefore, in the general case of a nonuniform anisotropic medium and arbitrary dislocation, the effect of each element of it with respect to the displacement caused by it will be equivalent to the effect of three couples without moment.

Mutual orientation of the major axes of the moment tensor of equivalent dipoles and vectors n_i and a_j, which determine the dislocation element, depend on the anisotropy of the medium and can be investigated only for particular cases. The most important is the case of an isotropic medium, where, from (1.3.17) and (3.2.13), we get

$$m_{lm} = \lambda \delta_{lm} a_{(n)} + \mu (a_l n_m + a_m n_l) \qquad (3.2.17)$$

Let us resolve the vector a_j into two components, the tensile component $a_{(n)} n_j$ and slip component a_j'', tangent to the dislocation surface and consider the corresponding tensors m_{lm} separately. Let us choose the system of coordinates in such a way that a_j'' is directed along the x_1 axis and n_j along the x_2 axis. Then, if the tensile component is equal to zero, $a_{(n)} = 0$, only the components $m_{12} = \mu a''$ and $m_{21} = \mu a''$ will be nonzero, where a'' is the modulus of a_j'', that is, the magnitude of slip. So we obtain the situation shown in Figure 3.1, where the $x_1 x_3$ plane is the dislocation plane and the slip is directed along the x_1 axis. It is seen that the x_2 axis is one of the principal axes of the tensor M_{lm}, which is perpendicular to the plane containing the normal to the dislocation and slip vector, whereas the other two principal axes are at 45° to the x_1 and x_2 axes. In the case of pure tensile dislocation, $a_j'' = 0$, choosing the x_1 and x_3 axes arbitrarily in the dislocation plane and the x_2 axis normal to it, we see that the nondiagonal components of tensor vanish, and the diagonal components are equal to

$$m_{11} = \lambda a_{(n)}; \qquad m_{22} = (\lambda + 2\mu) a_{(n)}; \qquad m_{33} = \lambda a_{(n)}$$

That is, the element of tensile dislocation is equivalent to a center of dilatation having an intensity of $(\lambda + 2\mu) a_{(n)}$ and to two concentrated couples without moment with intensities $2\mu a_{(n)}$, which are perpendicular to each other and lie in the plane of dislocation. A more general case can be studied where opening and slip occur simultaneously, but this case is more complicated.

In the seismological literature, the principal axes of moment tensor are usually denoted as P, T, and B axes, P being the axis of maximum compression (P for pressure) and T the axis of minimum compression (T for tension). These axes may, but not necessarily must, coincide with the principal axes of the initial stress tensor σ_{ik}^0. If one assumes that the plane of dislocation coincides with the plane of maximum shear stress of the tensor σ_{ik}^0 and the direction of slip coincides with the direction of this

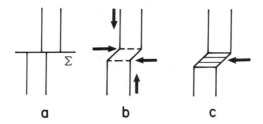

3.3 Equivalence of shear dislocation and elastic strain due to a concentrated dipole. (a) Shear dislocation; (b) elastic strain due to concentrated forces; (c) inelastic deformation, or flow (striped), equivalent to dislocation.

maximum shear stress, then the principal axes of m_{lm} and σ_{ik}^0 will coincide if there is no opening displacement. These two tensors, however, not only are quantitatively different but have a completely different nature, because σ_{ik}^0 represents the distribution of stresses actually present in the medium before the earthquake, whereas m_{lm} represents the distribution of fictitious dipoles that would create a similar elastic field in the medium in the absence of fracture.

Let us illustrate this by considering a column of the medium cut out along the normal to the dislocation. Due to fracture, its upper part is displaced relative to the lower part by the value a'' (Figure 3.3a). If concentrated forces (Figure 3.3b) are applied to the unbroken column, the upper part of the column will move relative to the lower only outside a layer whose thickness is equal to the arm of the couple, where large elastic deformation would occur. If the thickness of this layer tends toward zero, we get the situation illustrated in Figure 3.3a, but in this case, infinite elastic strain would be produced in an infinitely thin layer. To support such strain, concentrated forces are required, but these forces are not needed if the column is broken. As we argued in Chapter 1, if the fracture actually represents a band of plastic or viscous flow, we obtain the situation shown in Figure 3.3c. In this case the deformation is distributed as in Figure 3.3b, but in the striped area it is inelastic and no forces are required to support it. In this case, too, the elastic dipole, to which the dislocation is equivalent, is not actually applied to the medium.

The general solution (3.2.11) of the dislocation problem makes it possible to formulate the inverse problem of earthquake source theory in the following general way: At each point of the earth's surface S_0 let the displacement $u_i(\mathbf{x}, t)$ due to the earthquake be known. Next let the area

Σ_1 where slip has occurred during the earthquake also be known. Since $\Sigma(t)$ lies in Σ_1, and outside $\Sigma(t)$ the slip a_i is equal to zero, integration in (3.2.11) can be extended to the entire surface Σ_1. Then the inverse problem will consist of solving for the slip distribution $a_i(\mathbf{x}, t)$ at each point of Σ_1 for each instant t. For this, we have a system of integral equations, which follow from (3.2.11), when $\mathbf{x}_1 \in S_0$:

$$\int_0^{t_1} dt \int_{\Sigma_1} K_{ik}(\mathbf{x}_1, \mathbf{x}, t_1, t) a_i(\mathbf{x}, t) \, dS = u_k(\mathbf{x}_1, t_1) \qquad \text{for} \quad \mathbf{x}_1 \in S_0$$

$$(3.2.18)$$

where the kernel K_{ik} is given by

$$K_{ik} = c_{ijlm}(\mathbf{x}) n_j(\mathbf{x}) \frac{\partial}{\partial x_m} U_{kl}(\mathbf{x}_1, \mathbf{x}, t_1 - t) \qquad (3.2.19)$$

Imagine that the system of equations (3.2.18) is already solved and $a_i(\mathbf{x}, t)$ is known on Σ_1. Then $\Sigma(t)$ determines the part of Σ_1 on which a_i is nonzero at time t. Next, substituting $a_i(\mathbf{x}, t)$ into (3.2.11), we would obtain the displacement u_k at every point outside the dislocation and for every instant. Now from Hooke's law, we can find the stress perturbation $\tau_{ij}(\mathbf{x}_1, t_1)$ at every point $\mathbf{x}_1$ outside the dislocation. Furthermore, by tending $\mathbf{x}_1$ toward the dislocation surface $\Sigma(t)$, we can find the stress drop history at every point of fracture. A complete mechanical description of the process at the earthquake source would then be obtained.

Thus, the inverse problem of the earthquake source theory, in general, consists of the solution of the system of integral equations (3.2.18). The source models considered in seismology (e.g., Archuleta, 1984) in which the slip history $a_i(\mathbf{x}, t)$ was assigned *a priori*, represent, essentially, a trial-and-error approach to the solution of the system (3.2.18). To evaluate the prospect of such attempts, it is necessary to investigate the solvability of the inverse problem, that is, the questions of the existence, uniqueness, and stability of the solution of the system of equations (3.2.18). To do this for the general case, one must have the expression for the kernel K_{ik} or, equivalently, the expression for Green's tensor U_{ik} for the whole earth. Unfortunately, even if one succeeded in obtaining this expression, it would hardly enable one to investigate the solvability of equation (3.2.18), since it would be too complicated. To make such an investigation, one must use the particular form of seismic information representation. Namely, a seismogram $u_k(\mathbf{x}_1, t_1)$ can be split into records of individual waves, body waves, and surface waves. Then one must also split the kernel K_{ik} into the sum of several terms corresponding to each separate wave and write the equation (3.2.18) for each wave separately. Thus, several independent problems are obtained for the determination of $a_i(\mathbf{x}, t)$ from different waves. For body waves, there is an additional simplification because, in most cases, they can be described by the first ray approximation. This is possible when the pulse duration is much shorter than the travel time, that is, when the observations are made at sufficiently large epicentral distances. Let us assume that the source can be surrounded by a closed surface within which the medium can be considered to be homogeneous. Then, using ray theory, one can recalculate the displacement distribution from the earth's surface onto this surface and then evaluate the elastic field replacing the earth outside this surface by an infinite, homogeneous medium. This procedure (which includes corrections for reflections from the earth's free surface near the source region) is called straightening of rays. In what follows, we will not consider this question, assuming that the rays are already straightened – that is, that the medium is infinite and homogeneous.

3.3 Formulation of the inverse problem for far-field body waves

Let us assume that the observations have already been reduced to a uniform medium with density and elastic constants equal to those in the neighborhood of the source. The equations for the inverse problem for such a case can be obtained if we use the expression for Green's tensor for a homogeneous medium in equations (3.2.19) and the simplifications based on the assumption that the body waves can be described by the first ray approximation. The expression for Green's tensor for a homogeneous, isotropic medium is well known (Stokes' solution; see, e.g., Aki and Richards, 1980). In our notation, it is

$$
U_{ik}(\mathbf{x}, \mathbf{x}_0, t - t_0)
$$

$$
= \frac{1}{4\pi\mu} \left\{ v_S^2 \frac{\partial^2}{\partial x_i \, \partial x_k} \left(\frac{t - r/v_P - t_0}{r} H\left(t - \frac{r}{v_P} - t_0 \right) \right) \right.
$$

$$
+ \left[\delta_{ik} \frac{\delta(t - r/v_S - t_0)}{r} - v_S^2 \frac{\partial^2}{\partial x_i \, \partial x_k} \right.
$$

$$
\left. \left. \times \left(\frac{t - r/v_S - t_0}{r} H\left(t - \frac{r}{v_S} - t_0 \right) \right) \right] \right\} \qquad (3.3.1)
$$

where

$$r = \sqrt{(x_i - x_{i0})(x_i - x_{i0})}$$

is the distance from the source $\mathbf{x}_0$ to the observation point $\mathbf{x}$, and $H(t)$ is the Heaviside step function. In this expression, the first term in braces represents the perturbation propagating with the longitudinal wave velocity v_P (i.e., the longitudinal wave) and the second term represents the transverse wave. Let us introduce separate notations for longitudinal and transverse perturbations:

$$U_{ik}^P(\mathbf{x}, \mathbf{x}_0, t - t_0) = \frac{v_S^2}{4\pi\mu} \frac{\partial^2}{\partial x_i \partial x_k} \left(\frac{t - r/v_P - t_0}{r} H\left(t - \frac{r}{v_P} - t_0 \right) \right)$$

$$(3.3.2)$$

$$U_{ik}^S(\mathbf{x}, \mathbf{x}_0, t - t_0) = \frac{1}{4\pi\mu} \left[\delta_{ik} \frac{\delta(t - r/v_S - t_0)}{r} \right.$$

$$\left. - v_S^2 \frac{\partial^2}{\partial x_i \partial x_k} \left(\frac{t - r/v_S - t_0}{r} H\left(t - \frac{r}{v_S} - t_0 \right) \right) \right]$$

$$(3.3.3)$$

Now the displacements for longitudinal and transverse waves radiated from the dislocation can be obtained by substituting expressions (3.3.2) and (3.3.3) into (3.2.14):

$$u_k^P(\mathbf{x}_1, t_1) = \int_0^{t_1} dt \int_{\Sigma(t)} m_{lm}(\mathbf{x}, t) \frac{v_S^2}{4\pi\mu} \frac{\partial^3}{\partial x_k \partial x_l \partial x_m}$$

$$\times \left(\frac{t_1 - r/v_P - t}{r} H\left(t_1 - \frac{r}{v_P} - t \right) \right) dS \qquad (3.3.4)$$

$$u_k^S(\mathbf{x}_1, t_1) = \int_0^{t_1} \int_{\Sigma(t)} m_{lm}(\mathbf{x}, t) \frac{1}{4\pi\mu} \frac{\partial}{\partial x_m} \left[\frac{\delta_{kl} \delta(t_1 - r/v_S - t)}{r} \right.$$

$$\left. - v_S^2 \frac{\partial^2}{\partial x_k \partial x_l} \left(\frac{t_1 - r/v_S - t}{r} H\left(t_1 - \frac{r}{v_S} - t \right) \right) \right] dS \, dt \quad (3.3.5)$$

where $r = \{(x_i - x_{i1})(x_i - x_{i1})\}^{1/2}$ and the moment tensor density m_{lm} is given by (3.2.17). Obviously, u_k^P and u_k^S satisfy the corresponding wave equations:

$$u_{k,jj}^P = \frac{1}{v_P^2} \ddot{u}_k^P; \qquad u_{k,jj}^S = \frac{1}{v_S^2} \ddot{u}_k^S \qquad (3.3.6)$$

Since the derivatives with respect to the coordinates $\mathbf{x}$ and $\mathbf{x}_1$ differ only in sign, using these relations, we can write equations (3.3.4) and (3.3.5) as

$$u_k^P(\mathbf{x}_1, t_1) = -\frac{1}{4\pi\rho} \frac{\partial^3}{\partial x_{k1} \partial x_{m1} \partial x_{l1}}$$

$$\times \int_{\Sigma_1} \int_0^{t_1 - r/v_P} m_{lm}(\mathbf{x}, t) \frac{t_1 - r/v_P - t}{r} dt \, dS \qquad (3.3.7)$$

$$u_k^S(\mathbf{x}_1, t_1) = -\frac{1}{4\pi\rho} \frac{\partial}{\partial x_{m1}} \left(\delta_{kl} \frac{\partial^2}{\partial x_{j1} \partial x_{j1}} - \frac{\partial^2}{\partial x_{k1} \partial x_{l1}} \right)$$

$$\times \int_{\Sigma_1} \int_0^{t_1 - r/v_S} m_{lm}(\mathbf{x}, t) \frac{t_1 - r/v_S - t}{r} dt \, dS \qquad (3.3.8)$$

where the integration domain $\Sigma(t)$ is replaced, as in (3.2.19), by Σ_1 and the integrations over Σ_1 and t are interchanged.

Let us replace the integration variable t by $(t_1 - r/v - t)$. Then

$$u_k^P(\mathbf{x}_1, t_1) = -\frac{1}{4\pi\rho} \frac{\partial^3}{\partial x_{k1} \partial x_{m1} \partial x_{l1}}$$

$$\times \int_{\Sigma_1} \int_0^{t_1} m_{lm}\left(\mathbf{x}, t_1 - \frac{r}{v_P} - t\right) \frac{t}{r} dt \, dS \qquad (3.3.9)$$

$$u_k^S(\mathbf{x}_1, t_1) = -\frac{1}{4\pi\rho} \frac{\partial}{\partial x_{m1}} \left(\delta_{kl} \frac{\partial^2}{\partial x_{j1} \partial x_{j1}} - \frac{\partial^2}{\partial_{k1} \partial x_{l1}} \right)$$

$$\times \int_{\Sigma_1} \int_0^{t_1} m_{lm}\left(\mathbf{x}, t_1 - \frac{r}{v_S} - t\right) \frac{t}{r} dt \, dS \qquad (3.3.10)$$

The moment tensor density m_{lm} is proportional to the displacement jump a_i and, consequently, vanishes for negative time. Moreover, assuming for simplicity that the fracture propagation velocity does not exceed v_S, we have

$$m_{lm}(\mathbf{x}, t) = 0 \quad \text{for} \quad t < \frac{s}{v_S} \qquad (3.3.11)$$

where s is the distance of the point $\mathbf{x}$ from the fracture initiation point. Hence, the integrands in (3.3.9) and (3.3.10) are nonzero only for

$$0 < t < t_1 - \frac{r}{v_P} - \frac{s}{v_S}$$

and

$$0 < t < t_1 - \frac{r}{v_S} - \frac{s}{v_S}$$

respectively. Let R be the distance from the fracture initiation point to the observation point x_1. It is obvious that $r + s > R$. The right side of the preceding inequalities can be replaced by $t_1 - R/v_P$ and $t_1 - R/v_S$, respectively. Now it is seen that expressions (3.3.9) and (3.3.10) are nonvanishing only at $t_1 > R/v_P$ and $t_1 > R/v_S$, and these equations can be written as

$$u_k^P(x_1, t' + T_P) = -\frac{1}{4\pi\rho}\frac{\partial^3}{\partial x_{k1}\,\partial x_{l1}\,\partial x_{ml}}$$

$$\times \int_{\Sigma_1}\frac{1}{r}\int_0^{t'}m_{lm}\left(x, t' + \frac{R}{v_P} - \frac{r}{v_P} - t\right)t\,dt\,dS$$

(3.3.12a)

$$u_k^S(x_1, t' + T_S) = -\frac{1}{4\pi\rho}\frac{\partial}{\partial x_{ml}}\left(\delta_{kl}\frac{\partial^2}{\partial x_{j1}\,\partial x_{j1}} - \frac{\partial^2}{\partial x_{k1}\,\partial x_{l1}}\right)$$

$$\times \int_{\Sigma_1}\frac{1}{r}\int_0^{t'}m_{lm}\left(x, t' + \frac{R}{v_S} - \frac{r}{v_S} - t\right)t\,dt\,dS$$

(3.3.12b)

where t' is the time measured from the arrival time of the wave at the observation point and T_P and T_S are the travel times of the waves:

$$T_P = \frac{R}{v_P}; \qquad T_S = \frac{R}{v_S}$$

(3.3.13)

Thus, the first arrivals are due to the waves radiated from the fracture initiation point. Let us choose this point (the hypocenter) as the origin of the coordinate system. Then $R = \sqrt{x_{i1}x_{i1}}$. Now let us assume that the distance R is much greater than the maximum fracture dimension d (for a given R, this corresponds to sufficiently small t' and for fixed t' to sufficiently large R). Then we get the asymptotic equation

$$r = R - x_i m_i + O(d^2/R)$$

(3.3.14)

where m_i is the unit vector directed from the hypocenter to the observation point x_i:

$$m_i = x_{i1}/R$$

(3.3.15)

Now, to the accuracy of $O(d/R)$, we get from (3.3.12a) and (3.3.12b):

$$u_k^P(\mathbf{x}_1, t' + T_P) = \frac{-1}{4\pi\rho v_P^3 R} m_k m_l m_m \int_{\Sigma_1} \dot{m}_{lm}\left(\mathbf{x}, t' + \frac{m_i x_i}{v_P}\right) dS$$

(3.3.16)

$$u_k^S(\mathbf{x}_1, t' + T_P) = \frac{1}{4\pi\rho v_S^3 R}(m_k m_l - \delta_{kl}) m_m \int_{\Sigma_1} \dot{m}_{lm}\left(\mathbf{x}, t' + \frac{m_i x_i}{v_S}\right) dS$$

(3.3.17)

These equations hold for

$$\frac{v_S t'}{R} \ll 1$$

(3.3.18)

that is, in the zeroth ray approximation. We note that the displacement vector for the longitudinal wave u_k^P is collinear with the unit ray vector m_k. Its component along the ray u_P gives the magnitude of the P displacement, and its sign is positive when directed to the source. Therefore, the three equations of (3.3.16) are equivalent to one scalar equation:

$$u_P(\mathbf{x}_1, t' + T_P) = \frac{m_l m_m}{4\pi\rho v_P^3 R} \int_{\Sigma_1} \dot{m}_{lm}\left(\mathbf{x}, t' + \frac{m_i x_i}{v_P}\right) dS \quad (3.3.19)$$

It is now easy to obtain the corresponding formula for an inhomogeneous earth. In the ray approximation, we have

$$u_P(\mathbf{x}_1, t' + T_P) = \frac{A^P(t', \mathbf{m})}{R_P(\Delta, h)}$$

(3.3.20)

where $R_P(\Delta, h)$ is the geometric spreading factor for the epicentral distance Δ and depth h of the source, and A^P is the "amplitude" or pulse shape, which depends on the ray direction $\mathbf{m}$. Near the source, where the medium is assumed to be homogeneous, the rays are straight and the spreading factor R_P coincides with the distance R from the source. Therefore, in formula (3.3.19), Ru_P can be replaced by A^P and we obtain

$$m_l m_m \int_{\Sigma_1} \dot{m}_{lm}\left(\mathbf{x}, t + \frac{m_i x_i}{v_P}\right) dS = 4\pi\rho v_P^3 A^P(t, \mathbf{m}) \quad (3.3.21)$$

where m_i are the direction cosines of the tangent to the ray near the source, and A^P is obtained from the record of the longitudinal wave using (3.3.20), where R_P includes corrections for reflection and transmission at interfaces and at the earth's surface.

To obtain a similar relation for transverse waves, we take into account the fact that in the layered earth, the waves SH and SV propagate independently, and relations similar to (3.3.20) exist:

$$u_{SV1}(\mathbf{x}, t' + T_S) = \frac{A^{SV}(t', \mathbf{m})}{R_{SV}(\Delta, h)} \tag{3.3.22}$$

and

$$u_{SH1}(\mathbf{x}, t' + T_S) = \frac{A^{SH}(t', \mathbf{m})}{R_{SH}(\Delta, h)} \tag{3.3.23}$$

where R_{SV} and R_{SH} are the SV and SH spreading factors, A^{SV} and A^{SH} are the amplitudes, and u_{SH1} and u_{SV1} are the displacements in SH and SV waves at the observation point.

Let $\mathbf{m}_1$ and $\mathbf{z}_1$ be the unit vectors at the observation point, directed along the ray and along the vertical, respectively, and let $\mathbf{m}$ and $\mathbf{z}$ be similar vectors at the source. Then

$$u_{SH1} = \frac{u_{k1}^S e_{kij} m_{i1} z_{j1}}{\left(1 - (m_{r1} z_{r1})^2\right)^{1/2}}$$

$$\tag{3.3.24}$$

$$u_{SV1} = \frac{u_{k1}^S (z_{k1} - m_{k1} m_{j1} z_{j1})}{\left(1 - (m_{r1} z_{r1})^2\right)^{1/2}}$$

and

$$u_{SH} = \frac{u_k^S e_{kij} m_i z_j}{\left(1 - (m_r z_r)^2\right)^{1/2}}$$

$$\tag{3.3.25}$$

$$u_{SV} = \frac{u_k^S (z_k - m_k m_j z_j)}{\left(1 - (m_r z_r)^2\right)^{1/2}}$$

where u_{SH} and u_{SV} are the displacements in SH and SV waves at the source. Furthermore, we have

$$u_{SV} = \frac{A^{SV}}{R} = \frac{u_{SV1} R_{SV}}{R}$$

$$u_{SH} = \frac{A^{SH}}{R} = \frac{u_{SH1} R_{SH}}{R} \tag{3.3.26}$$

Equations (3.3.25) can be solved for u_k^S:

$$u_k^S = \frac{u_{SV}(z_k - m_k m_j z_j) + u_{SH} e_{kij} m_i z_j}{\left(1 - (m_r z_r)^2\right)^{1/2}} \qquad (3.3.27)$$

Using (3.3.24) we get

$$Ru_k^S(\mathbf{x}, t' + T_S) \equiv A_k^S(t', \mathbf{m})$$

$$= \frac{(z_k - m_k m_j z_j) A^{SV}(t', \mathbf{m}) + e_{kij} m_i z_j A^{SH}(t', \mathbf{m})}{\left(1 - (m_r z_r)^2\right)^{1/2}}$$

$$(3.3.28)$$

Putting this expression into (3.3.17), we finally get

$$(m_k m_l - \delta_{kl}) m_m \int_{\Sigma_1} \dot{m}_{lm}\left(\mathbf{x}, t - \frac{m_i x_i}{v_S}\right) dS = 4\pi \rho v_S^3 A_k^S(t, \mathbf{m}) \quad (3.3.29)$$

It can be seen from (3.3.28) and (3.3.29) that both the left and right sides are orthogonal to the ray vector $\mathbf{m}$; that is, only two of the three equations of (3.3.29) are independent. The right side in (3.3.29) is calculated from the records of u_{SV} and u_{SH} of waves SV and SH at the point of observation according to formulas (3.3.28), (3.3.22), and (3.3.23), and as for P waves, the spreading factors must include corrections for reflection and transmission at the interfaces. Similar relations when the fracture propagation speed exceeds v_S can be obtained mutatis mutandis.

Equations (3.3.21) and (3.3.29) comprise the formulation of the inverse problem of source theory for far-field body waves. If the surface Σ_1 along which the fracture propagates is known, the equation (3.2.17) for $m_{lm}(\mathbf{x}, t)$ shows that (3.3.21) and (3.3.29) will contain only three unknown functions of $a_i(\mathbf{x}, t)$, that is, as many as the number of equations. Before attempting a solution of equations (3.3.21) and (3.3.29), we must determine the position and shape of the fracture surface at the source. It is usually assumed that the fracture surface Σ is a plane. In this case, it is sufficient to determine its orientation. Near the first arrivals, only a small part of the fracture surface in the vicinity of the initiation point contributes to the seismogram. Therefore, for the first arrivals it is justifiable to assume, as usual, that the fracture surface is planar.

In the following section, we will consider the possibility of determining the position of the fault plane directly from equations (3.3.21) and (3.3.29). It should be noted here, however, that there is another and perhaps more effective method of determining the fault plane. This method uses the tendency of the fracture process to be discrete. As mentioned in Chapter 2, the fracture process is "microscopically" discrete and consists of nucleation and coalescence of microfractures. Sometimes it is possible to identify on seismograms the arrivals that are associated with such stops and subsequent starts of the fracture propagation. These are called stopping phases. Hypocenter locations for these phases give the positions of these stops. Since these points lie on Σ_1, with a sufficient number of these points, it is possible to determine the shape and position of the fracture surface and the average fracture velocity between the stops. Such an approach was considered by Boatwright (1980, 1981) and Cichowicz (1981).

The discreteness of fracture propagation is manifested in the discreteness of the radiation of seismic waves. Therefore, the application of this method depends on how the seismic signal was smoothed. This smoothing occurs due to the attenuation of high frequencies during propagation as well as to the limited frequency band of seismic instruments. Equations (3.3.21) and (3.3.29) are directly applicable only to records on which sharp variations in radiation are smoothed out, that is, for wavelengths and periods longer than the space–time scale of radiation discreteness. We will call this coherent radiation.

At the other extreme, individual discrete pulses may be considered independent and uncorrelated in time at such short periods. We will call this incoherent radiation. In the case of incoherent radiation, formulas (3.3.21) and (3.3.29) are not applicable to the source as a whole, but they can be applied to the radiation of each of the microfractures constituting the macrofracture. On this basis, it is possible to obtain from these equations other relations that are applicable to incoherent radiation.

Teleseismic observations, especially at long periods, probably provide mainly the coherent part of source radiation. Short-period records in the near field and seismic intensity seem to be determined primarily by incoherent radiation.

We made the assumption about the coherentness, in fact, when going from equations (3.3.12a) and (3.3.12b) to equations (3.3.16) and (3.3.17). Namely, we used the asymptotic equations

$$
m_{lm}\left(\mathbf{x}, t' - t + \frac{x_i m_i}{v_{\mathrm{P}}} + O\left(\frac{d^2}{v_{\mathrm{P}} R}\right)\right)
$$

$$
= m_{lm}\left(\mathbf{x}, t' - t + \frac{x_i m_i}{v_{\mathrm{P}}}\right) + O\left(\frac{d^2}{R v_{\mathrm{P}}}\right) \tag{3.3.30}
$$

This is true only if $m_{lm}(\mathbf{x}, t)$ and consequently $a_i(\mathbf{x}, t)$ are assumed to be sufficiently smooth.

Without making this assumption, instead of (3.3.16) and (3.3.17) we would obtain the following relations (again omitting the terms of order d/R):

$$u_k^{\mathrm{P}}(\mathbf{x}_1, t_1) = -\frac{m_k m_l m_m}{4\pi \rho v_{\mathrm{P}}^3 R} \int_{\Sigma_1} \dot{m}_{lm}\left(\mathbf{x}, t_1 - \frac{r}{v_{\mathrm{P}}}\right) dS \qquad (3.3.31)$$

$$u_k^{\mathrm{S}}(\mathbf{x}_1, t_1) = \frac{(m_k m_l - \delta_{kl}) m_m}{4\pi \rho v_{\mathrm{S}}^3 R} \int_{\Sigma_1} \dot{m}_{lm}\left(\mathbf{x}, t_1 - \frac{r}{v_{\mathrm{S}}}\right) dS \qquad (3.3.32)$$

Now for an inhomogeneous medium, R must be replaced by the corresponding spreading factors, r/v_{P} or r/v_{S} by $T_{\mathrm{P}}(\mathbf{x}, \mathbf{x}_1)$ or $T_{\mathrm{S}}(\mathbf{x}, \mathbf{x}_1)$, which depend on the position on the fracture surface, and the dependence of displacement direction on the ray vector must be taken into account. Then we get

$$u^{\mathrm{P}}(\mathbf{x}_1, t_1) = \frac{m_l m_m}{4\pi \rho v_{\mathrm{P}}^3 R_{\mathrm{P}}(\Delta, h)} \int_{\Sigma_1} \dot{m}_{lm}(\mathbf{x}, t_1 - T_{\mathrm{P}}(\mathbf{x}, \mathbf{x}_1)) \, dS \qquad (3.3.33)$$

$$u^{\mathrm{SV}}(\mathbf{x}_1, t_1) = \frac{(m_k m_l - \delta_{kl}) z_k m_m}{4\pi \rho v_{\mathrm{S}}^3 R_{\mathrm{SV}}(\Delta, h)\left(1 - (m_r z_r)^2\right)^{1/2}}$$

$$\times \int_{\Sigma_1} \dot{m}_{lm}(\mathbf{x}, t_1 - T_{\mathrm{S}}(\mathbf{x}, \mathbf{x}_1)) \, dS \qquad (3.3.34)$$

$$u^{\mathrm{SH}}(\mathbf{x}_1, t_1) = \frac{e_{lij} z_i m_j m_m}{4\pi \rho v_{\mathrm{S}}^3 R_{\mathrm{SH}}(\Delta, h)\left(1 - (m_r z_r)^2\right)^{1/2}}$$

$$\times \int_{\Sigma_1} \dot{m}_{lm}(\mathbf{x}, t_1 - T_{\mathrm{S}}(\mathbf{x}, \mathbf{x}_1)) \, dS \qquad (3.3.35)$$

where z_k is the unit vertical vector at the source.

These formulas differ from the ones obtained using the assumptions (3.3.30) in that they contain the exact travel times $T_{\mathrm{P}}(\mathbf{x}, \mathbf{x}_1)$ or $T_{\mathrm{S}}(\mathbf{x}, \mathbf{x}_1)$, whereas in equations (3.3.21) and (3.3.29) each of these was replaced by approximate values

$$T_{\mathrm{P}}(\mathbf{x}, \mathbf{x}_1) \approx T_{\mathrm{P}}(0, \mathbf{x}_1) - \frac{m_i x_i}{v_{\mathrm{P}}}$$

$$T_{\mathrm{S}}(\mathbf{x}, \mathbf{x}_1) \approx T_{\mathrm{S}}(0, \mathbf{x}_1) - \frac{m_i x_i}{v_{\mathrm{S}}} \qquad (3.3.36)$$

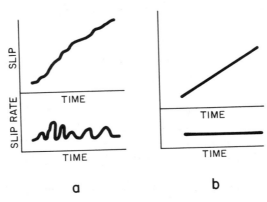

3.4   Influence of nonuniform motion on source radiation. (a) Slip and slip rate along a rough fault; (b) the same for a smooth fault.

Errors in these approximate formulas are of the order of

$$\Delta T_{\mathrm{P}} = O\left(\frac{d}{v_{\mathrm{P}}}\frac{d}{R}\right)$$

where $d = \max\sqrt{x_i x_i}$ is the distance from the initiation point to the farthest point of fracture and R is the hypocentral distance. This relation can be rewritten in another form as

$$\Delta T_{\mathrm{P}} = O\left(\frac{d^2}{v_{\mathrm{P}}^2 T_{\mathrm{P}}}\right) \qquad (3.3.37)$$

and similarly for transverse waves.

To find the conditions when errors of this order can be neglected, let us examine in more detail the process of radiation of waves by the fracture. Not only is the propagation of the fracture edge a discrete process, but so also is the slip along the fault. In fact, owing to the roughness of the fracture surface, individual asperities may catch one another during the slip, and the slip rate at these places might then decrease, if not vanish. This would continue until the asperities yielded, and the slip rate would then instantaneously increase. The curves in Figure 3.4a schematically represent the slip history at a point on the fault and its time derivative, that is, slip rate. Owing to the roughness of the fault material, slip occurs nonuniformly in time, and the slip rate plot looks like a sequence of pulses. If the process of slip is smoothed, a curve is obtained (Figure 3.4b). In this case, the slip rate is represented by a smooth curve without any peaks.

The integrands in (3.3.33) through (3.3.35) are proportional to the slip rate $\dot{a}$. Therefore, the radiation from every fault element is a sequence of pulses superimposed on a smooth function of time. Let the two pulses radiated by two fault elements have a duration of the order τ. The result of their superposition at the observation point depends on the difference of arrival times. If this delay is much smaller than τ, these pulses coalesce into one larger pulse. If the delay is greater than τ, these pulses will be recorded separately. The error ΔT_P does not affect the result of superposition of the pulses only if it is much smaller than the pulse duration τ. Thus, the error ΔT_P can be neglected if $(\Delta T_\mathrm{P}/\tau) \ll 1$, that is, when

$$\frac{d^2}{v_\mathrm{P}^2 T_\mathrm{P}\tau} \ll 1 \tag{3.3.38}$$

and similarly for ΔT_S. This condition depends on the pulse duration τ, and it might therefore be satisfied for some pulses and not satisfied for shorter ones. However, the width of the pulses recorded at the observation point depends not only on the duration of the radiated pulse but also on the filtering effect of the medium and the instrument. The effect of the medium and the instrument as linear filters can be expressed by an operator $\int' K(t - \tau)f(\tau)\,d\tau$, where $K(t)$ is the impulse response of the filter. Let the duration of $K(t)$ be τ_0. Then all the input pulses with width $\tau < \tau_0$ will be recorded as pulses with a width that is not less than τ_0. Therefore, for observations with a given instrument, the duration of all the pulses might be taken as greater than τ_0, and we get the condition

$$\frac{d^2}{v_\mathrm{P}^2 T_\mathrm{P}\tau_0} \ll 1 \tag{3.3.39}$$

The value τ_0 can be interpreted as the shortest period recorded under the given conditions, and (3.3.39) can be rewritten in the form

$$d^2 \ll R\lambda_{\min} \tag{3.3.40}$$

where R is the hypocentral distance and $\lambda_{\min}$ is the minimum wavelength recorded.

Thus, to replace the travel times by approximate values (3.3.36), the square of the largest fracture dimension must be much smaller than the product of hypocentral distance and the minimum wavelength recorded. Consequently, equations (3.3.21) and (3.3.29) are also applicable to incoherent radiation under the conditions

$$d \ll R; \qquad \frac{d}{R} \ll \frac{\lambda_{\min}}{d} \tag{3.3.41}$$

which have the same form for longitudinal and transverse waves.

The maximum fracture dimension depends on time and increases with fracture propagation. Therefore, conditions (3.3.41) are always satisfied at times sufficiently close to first arrivals.

3.4   First arrivals of body waves

Consider equations (3.3.21) and (3.3.29), which describe the radiation of body waves due to a fracture, for small values of t, that is, near the first arrival. In this case, the integrands are nonvanishing only for a small region around the initiation point. Assuming the fracture surface to be smooth, it can be replaced in this region by a part of its tangent plane. Then the normal n_k to the fracture surface can be considered constant (independent of position $\mathbf{x}$ on the surface) and can be taken outside the integral sign. In just the same way, because of the small integration area, the direction of the slip vector $a_i(\mathbf{x}, t)$ can be assumed to be independent of $\mathbf{x}$ and t,

$$a_i(\mathbf{x}, t) = b_i a(\mathbf{x}, t); \qquad b_i b_i = 1 \tag{3.4.1}$$

where b_i is the unit vector independent of $\mathbf{x}$ and t and $a(\mathbf{x}, t)$ is the magnitude of slip. Then the moment density (3.2.17) can be written as

$$m_{lm} = \{ \lambda \delta_{lm} n_k b_k + \mu(n_l b_m + n_m b_l) \} a(\mathbf{x}, t) \tag{3.4.2}$$

where the tensor in braces is independent of $\mathbf{x}$ and t, since λ and μ can likewise be replaced by their values at the origin of coordinates, that is, at the initiation point. Substituting (3.4.2) into equations (3.3.21) and (3.3.29), we obtain

$$\{ \lambda n_j b_j + 2\mu n_l m_l b_j m_j \} \int_{\Sigma_1} \dot{a}\left(\mathbf{x}, t + \frac{x_i m_i}{v_P}\right) dS$$

$$= 4\pi \rho v_P^3 A^P(t, \mathbf{m})$$

$$\mu \{ 2m_k(n_l m_l)(b_j m_j) - n_k(b_j m_j) - b_k(n_l m_l) \} \tag{3.4.3}$$

$$\times \int_{\Sigma_1} \dot{a}\left(\mathbf{x}, t + \frac{x_i m_i}{v_S}\right) dS = 4\pi \rho v_S^3 A_k^S(t, \mathbf{m})$$

Since, by definition, $a \geq 0$ and, for sufficiently small t, increases from zero, the integrals in these expressions are positive. Thus, the signs (polarization) of the first arrivals of body waves are determined only by the factors outside the integrals. These factors depend on the direction of the ray **m** and the two unit vectors **n** and **b** directed along the normal to the fracture plane and along the direction of slip. It can easily be seen that the expressions do not change during permutation of vectors **n** and **b**. Therefore, using the body wave polarization one cannot distinguish between **n** and **b**.

Let us now examine in detail the distribution of the signs of P waves. As has been observed in actual earthquakes, the rays corresponding to the different arrival signs of P waves are separated by a pair of planes (nodal planes), which are mutually orthogonal. The first equation of (3.4.3) shows that this is possible only in the case where the normal component of displacement jump vanishes, that is, $n_j b_j = 0$.

In this case, the sign of the P-wave displacement would be determined by the sign of the expression $(n_j m_j)(m_j b_j)$, and the equations of the nodal planes can be obtained by equating this expression with zero, that is,

$$n_i m_i = 0 \quad \text{or} \quad b_i m_i = 0 \tag{3.4.4}$$

The first of these is the equation of the plane normal to the vector **n**, that is, the fault plane, and the second is that of the plane normal to the slip vector *b*. In general, (when $\mathbf{nb} \neq 0$), the nodal surface is a hyperbolic cone whose axis is orthogonal to vectors **n** and **b**. Within observational errors, this cone is indistinguishable from a pair of orthogonal planes, if the normal component of the displacement jump is negligible.

The pair of nodal planes or the pair of orthogonal vectors **n** and **b** is called the fault-plane solution, first obtained by Vvedenskaya (1956). It is generally assumed that these vectors give the orientation of the concentrated dipole equivalent of the source. Equations (3.4.3) differ from the equations that would be obtained for a concentrated dipole only in that for a point source the integrals over Σ_1 will be replaced by time functions that are independent of the ray direction, that is, the source time function. Thus, the radiation due to the fracture is actually equivalent to the radiation of a point source, but only with respect to the distribution of the signs of first arrivals.

Let us denote

$$A_0^P(\mathbf{m}) = \lambda n_j b_j + 2\mu n_i m_i b_j m_j$$
$$A_{0k}^S(\mathbf{m}) = \mu \left\{ 2m_k (n_l m_l)(b_j m_j) - n_k (b_j m_j) - b_k (n_l m_l) \right\} \tag{3.4.5}$$

These functions describe the directivity of radiation of the point source equivalent to the source and can be calculated if the fault-plane solution is known. Thus, dividing each of the equations of the system (3.4.3) by the corresponding expression of (3.4.5), we get two functions:

$$
\int_{\Sigma_1} \dot{a}\left(\mathbf{x}, t + \frac{x_i m_i}{v_P}\right) dS = F_P(t, \mathbf{m})
$$

$$
\int_{\Sigma_1} \dot{a}\left(\mathbf{x}, t + \frac{x_i m_i}{v_S}\right) dS = F_S(t, \mathbf{m})
$$

(3.4.6)

whose dependence on the ray direction determines the ratio of the actual source radiation to that of the concentrated dipole. The relation of F_P and F_S to experimentally determined functions is given by the equations

$$
F_P(t, \mathbf{m}) = \frac{4\pi\rho v_P^3 A^P(t, \mathbf{m})}{A_0^P(\mathbf{m})}
$$

$$
F_S(t, \mathbf{m}) = \frac{4\pi\rho v_S^3 A_k^S(t, \mathbf{m})}{A_{0k}^S(\mathbf{m})}
$$

(3.4.7)

(where the summation is not to be carried out with respect to k). Here the functions A^P and A_k^S are related to the records of P, SV, and SH waves by the relations (3.3.20), (3.3.22), (3.3.23), and (3.3.28). Therefore, the functions (3.4.7) can be calculated for each station (choice of which fixes the direction $\mathbf{m}$ of the "straightened" ray), if the travel times and spreading factors are known. (Corrections for distortion of the signal by the instrument and the medium must be made.) In other words, the right sides of equations (3.4.6) should be considered to be experimentally determined functions.

Note now that the time dependence is not affected in the evaluation of (3.4.7) (not to mention the corrections for attenuation and distortion of the signal due to propagation). This indicates that F_P and F_S have the same shape as the records of P and S waves, respectively.

To investigate the possibility of determining the actual fault plane, consider the dependence of expressions (3.4.6) on the ray direction $\mathbf{m}$. First of all, let us note that the vector $\mathbf{x}$ belongs to the fault plane, and consequently it is orthogonal to the normal vector $\mathbf{n}$ of this plane, that is, $x_i n_i = 0$. The ray vector $\mathbf{m}$ can be resolved into two components, the first being orthogonal to the fault plane ($m_i^\perp$) and the second lying in this plane ($m_i^\parallel$):

$$
m_i = m_i^\perp + m_i^\parallel
$$

(3.4.8)

where

$$m_i^{\perp} = n_i m_k n_k; \qquad m_i^{\parallel} = m_i - n_i m_k n_k \qquad (3.4.9)$$

Obviously, $x_i m_i = x_i m_i^{\parallel}$; that is, the functions (3.4.6) depend only on the ray vector component parallel to the fault plane.

Assume that we have a record of a P wave at a point corresponding to the ray direction m_P at the source. Consider the S-wave record at another point corresponding to the ray direction m_S, the vectors m_P and m_S being related by

$$\frac{m_{Pi}^{\parallel}}{v_P} = \frac{m_{Si}^{\parallel}}{v_S} \qquad (3.4.10)$$

For a given m_P, there is a unique solution of (3.4.10) for m_S (since m_S and m_P are, by definition, unit vectors). It is now seen that under conditions (3.4.10), the left sides of equations (3.4.6), evaluated for directions m_P and m_S, respectively, would coincide, that is,

$$F_P(t, m_P) = F_S(t, m_S) \qquad (3.4.11)$$

In other words, the pulse shape of a P wave radiated in direction m_P coincides with the pulse shape of the S wave radiated in direction m_S, where m_S is related to m_P by equation (3.4.10).

Suppose we could select two stations such that the P-wave record at one of them coincides (after correction for signal distortion due to propagation) with the S-wave record at the other station. Let the directions of the rays (at the source) to these stations be m_P and m_S. Then these rays will be related to each other by conditions (3.4.10), and the normal vector to the fault will be determined from (3.4.10) as follows. Using the expression (3.4.9), rewrite the relation (3.4.10) as

$$\left(\frac{m_{Pk} n_k}{v_P} - \frac{m_{Sk} n_k}{v_S} \right) n_i = \frac{m_{Pi}}{v_P} - \frac{m_{Si}}{v_S} \qquad (3.4.12)$$

We see that the vector n is collinear with the vector on the right side

$$n_i = c \left(\frac{m_{Pi}}{v_P} - \frac{m_{Si}}{v_S} \right) \qquad (3.4.13)$$

where the constant c is found from the normalization condition $n_i n_i = 1$, which finally gives

$$n_i = \pm \frac{m_{Pi}/v_P - m_{Si}/v_S}{\left(1/v_P^2 + 1/v_S^2 - 2 m_{Pk} m_{Sk}/(v_P v_S) \right)^{1/2}} \qquad (3.4.14)$$

Considering equations (3.3.21) and (3.3.29), it is easily seen that the above arguments can be repeated not only for the neighborhood of first arrivals but also for the whole P and S waveforms under the condition that the fault is planar.

Thus, we obtain the method of the actual fault-plane determination from body wave records. This method is conceptually attractive in that it does not depend on any assumptions regarding the nature of motion at the source. The disadvantage of this method is the necessity of comparing pulse shapes at different stations; this requires accounting for signal distortion due to propagation, which is different for P and S waves. The S-wave records in actual seismograms are always of lower frequency than the P-wave records, and this shows that high frequencies in S waves are more attenuated than those in P waves. But this method requires the use of waves of sufficiently high frequency, because phase shifts $m_i x_i / v_P$ and $m_i x_i / v_S$ at low frequencies may be insignificant and the pulse shape would be the same for all stations (after accounting for propagation effects). However, the attenuation properties of the medium at high frequencies have not yet been investigated sufficiently to meet the requirements of this method.

To obtain a method without these drawbacks, it is necessary to make some more rigid assumptions regarding the slip history on the fault. As mentioned before, the first arrivals of body waves are related to fracture initiation. Let us assume that the initial stress σ_{ij}^0 in a sufficiently small region around the initiation point can be taken to be approximately homogeneous and the fracture propagation velocity during sufficiently short time to be approximately constant (possibly dependent on the propagation direction). Then, since the equations of motion in displacement are homogeneous in terms of space and time derivatives, the corresponding solution will be self-similar in terms of particle velocities and stresses; that is, the components of stress and particle velocity will be homogeneous functions of zero order in $\mathbf{x}$ and t. In particular, this is true for the slip rate $\dot{a}$, that is,

$$\dot{a}(\mathbf{x}, t) = f(\mathbf{x}/t) \qquad (3.4.15)$$

We substitute this expression into (3.4.6) to obtain

$$F_P(t, \mathbf{m}) = \int_{\Sigma_1} f\left(\frac{\mathbf{x}}{t + x_i m_i / v_P} \right) dS$$

$$F_S(t, \mathbf{m}) = \int_{\Sigma_1} f\left(\frac{\mathbf{x}}{t + x_i m_i / v_S} \right) dS \qquad (3.4.16)$$

By changing the integration variable, we can rewrite this expression as

$$F_P(t, \mathbf{m}) = t^2 \int_{\Sigma_1} f(\boldsymbol{\xi}) \frac{dS}{(1 - \xi_i m_i / v_P)^3}$$

$$F_S(t, \mathbf{m}) = t^2 \int_{\Sigma_1} f(\boldsymbol{\xi}) \frac{dS}{(1 - \xi_i m_i / v_S)^3}$$

(3.4.17)

The integrations in these expressions are carried out over the part of the plane Σ_1 on which $f(\boldsymbol{\xi}) > 0$. From (3.4.15), it can be seen that $|\boldsymbol{\xi}|$ does not exceed the maximum propagation velocity $v_{\max}$. It was assumed above that $v_{\max} < v_S$. For this case we get the inequality

$$1 - \frac{\xi_i m_i}{v_S} > 1 - \frac{|\boldsymbol{\xi}|}{v_S} > 1 - \frac{v_{\max}}{v_S} > 0$$

and, moreover, $1 - \xi_i m_i / v_P > 0$. Thus, the integrands in (3.4.17) are positive for all the values of $\mathbf{m}$. Assume now that the propagation velocity is close to v_S at least in one direction (say, $0.8 v_S$). Then the main contribution to the values of the integrals in (3.4.17) will be from the neighborhood of the fracture edge, since the expressions in the denominator will be, at suitable $\mathbf{m}$, several times greater near the edge than at $\boldsymbol{\xi} = 0$ (i.e., at the initiation point). This conclusion becomes even stronger if we note (as discussed in Chapter 2) that the slip rate $\dot{a}$ has a square root singularity at the edge. Accurate analysis is possible only for particular forms of $f(\boldsymbol{\xi})$, but it is clear that the maximum values of (3.4.17) correspond to the rays for which the magnitude of

$$\sup_{\boldsymbol{\xi}} (\xi_i m_i)$$

has a maximum. Since ξ_i is on the fault plane, it follows that the maximum of the expressions (3.4.17) corresponds to $\mathbf{m}$ parallel to the fault plane and in that direction where $|\boldsymbol{\xi}|$ is maximum, that is, in the direction of the fastest fracture propagation.

This conclusion has a simple physical meaning. To understand it, consider the cross section of the source by a plane normal to the fault surface (the plane of the paper in Figure 3.5a). The propagation of fracture can be considered to be a sequence of concentrated impulsive sources $s_1, s_2, \ldots$, acting at instants $t_1, t_2, \ldots$, respectively. For the sake of simplicity, we consider only the radiation of P waves. Then we can obtain the radiation pattern, which depends on the fracture velocity, by drawing the wavelet fronts radiated by each elementary source (Benioff, 1955).

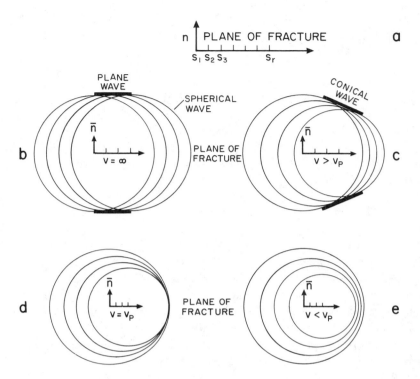

3.5   Effect of fracture velocity on radiation in different directions. (a) Plane of fracture; (b) instantaneous fracture; (c) fracture speed higher than the P-wave speed; (d) fracture speed equal to the P-wave speed; (e) fracture speed less than the P-wave speed.

First, let the fracture occur instantaneously. Then all the elementary sources act simultaneously and the wave fronts shown in Figure 3.5b are obtained. Every elementary source radiation is not spherically symmetric, but the directivity of far-field radiation of all the sources is the same [as given by (3.4.5)] and it can be eliminated as in (3.4.6) and (3.4.16). Examining Figure 3.5b, we see that (after elimination of the point source directivity), radiation from the instantaneous fracture is equivalent, in the far field, to a spherical wave whose intensity (which is roughly proportional to the density of the wavelets) depends on the direction. It is minimum along the fault plane and increases toward its normal, where two plane waves propagate away from the fault plane. Because these plane waves do not suffer geometric spreading whereas the amplitude of the spherical waves decreases proportionally to the distance from the source, we see that the radiation has a relative maximum in the direction of the normal to the fault. This maximum becomes sharper as the distance between the source and the observation point increases. In contrast, the duration of radiation is greater for the directions along the fault plane.

Instantaneous fracture implies infinite propagation velocity, which, as stated in Chapter 2, is impossible. Now let the fracture propagate at a finite velocity, higher than the P-wave velocity (which is also impossible). The corresponding pattern of wavelet fronts is shown in Figure 3.5c, where, for simplicity, unilateral propagation is assumed. Comparing this picture with the previous one, we see that the radiation maximum is not in the direction of the fault normal but in the direction that makes an angle $\theta = \cos^{-1}(v_P/v)$ with it, in which the conical front propagates (envelope of wavelet fronts). When the fracture velocity becomes equal to v_P, the pattern shown in Figure 3.5d is obtained. In this case, the radiation maximum (infinite in the far-field approximation) is in the direction of propagation, and the amplitude decreases in all directions, as for spherical waves. Finally, if the fracture velocity is less than the P-wave velocity (Figure 3.5e), the radiation maximum is in the direction of fracture propagation, but becomes weaker as the fracture velocity decreases. This last case is the only one that is physically possible.

To illustrate these conclusions, we shall present the solutions of two crack problems. For a self-similar, circular shear crack with constant stress drop $\Delta\sigma$, the slip rate distribution is (Kostrov, 1964)

$$\dot{a} = \frac{tA}{\left(t^2 - x_i x_i/v^2\right)^{1/2}} \qquad \text{for} \quad x_i x_i < t^2 v^2 \qquad (3.4.18)$$

Here, v is the constant fracture velocity, which is $\leq v_R$, the Rayleigh wave velocity, and A is given by

$$A = \frac{4\,\Delta\sigma}{\mu I(v)} \qquad (3.4.19)$$

where $I(v)$ is a smooth function of v given by

$$I(v) = \left(\frac{v^2}{v_S^3}\right)\int_0^\infty \left\{(2\nu + 1)^2 - 4\nu(\gamma^2 + \nu)^{1/2}(1 + \nu)^{1/2} + 1 + \nu\right\}$$

$$\times \frac{d\nu}{\left[1 + \nu(v^2/v_S^2)\right]^2(1 + \nu)^{1/2}} \qquad (3.4.20)$$

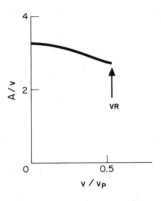

3.6 A/v, normalized by its static value, versus fracture speed v.

and $\gamma = v_S/v_P$.[1] Here A/v gives the ratio of slip versus crack radius and is plotted against v in Figure 3.6, its value being normalized by the static value $24\,\Delta\sigma/7\pi\mu$. Expression (3.4.18) has the form of (3.4.15), where

$$f(\boldsymbol{\xi}) = \frac{A}{\left(1 - \xi_i\xi_i/v^2\right)^{1/2}} \tag{3.4.21}$$

Integrals of (3.4.17) can now be evaluated in closed form, giving

$$F_P(t,\mathbf{m}) = \frac{2\pi A v^2 t^2}{\left(1 - \left(v^2/v_P^2\right)\cos^2\theta\right)^2}$$

$$F_S(t,\mathbf{m}) = \frac{2\pi A v^2 t^2}{\left(1 - \left(v^2/v_S^2\right)\cos^2\theta\right)^2} \tag{3.4.22}$$

where θ is the angle between the direction of the ray $\mathbf{m}$ and the fault plane, that is,

$$\cos\theta = \left(1 - \left(m_i n_i\right)^2\right)^{1/2} = m_i^{\parallel} \tag{3.4.23}$$

---

[1] Performing the integration in (3.4.20), we can write $I(v)$ in closed form as

$$I(v) = \alpha^4\left\{12(\gamma - 1) - \frac{12\alpha^4 - 21\alpha^2 + 8}{\left(\alpha^2 - 1\right)^{(3/2)}}\tan^{-1}\sqrt{\alpha^2 - 1}\right.$$

$$\left. + \frac{12\alpha^2 - 8\gamma^2}{\sqrt{\alpha^2 - \gamma^2}}\tan^{-1}\frac{\sqrt{\alpha^2 - \gamma^2}}{\gamma} + \frac{\alpha^2 - 2}{\alpha^2(\alpha^2 - 1)}\right\} \tag{3.4.20a}$$

where $\alpha^2 = v_S^2/v^2$.

As expected, the expressions (3.4.22) are maximum at $\theta = 0$, that is, along the fault plane. The value of this maximum increases with increasing crack propagation velocity v.

Burridge and Willis (1969) solved a more general problem, in which the fracture velocity depended on the direction, so that the crack at all times had an elliptical shape, given by

$$\frac{x_1^2}{v_1^2} + \frac{x_2^2}{v_2^2} \leq t^2 \tag{3.4.24}$$

where v_1 and v_2 are the fracture speeds in the x_1 and x_2 directions, respectively. It was found that the slip direction coincides with the direction of stress drop, when it is parallel to one of the axes of the ellipse. The distribution of slip rate magnitude is

$$\dot{a}(\mathbf{x}, t) = \frac{At}{\left(t^2 - x_1^2/v_1^2 - x_2^2/v_2^2\right)^{1/2}} \tag{3.4.25}$$

where A is proportional to the stress drop $\Delta\sigma$. The expression for A has the form of a complicated surface integral, which depends on the velocities v_1 and v_2.

In this case, $f(\boldsymbol{\xi})$ takes the form

$$f(\boldsymbol{\xi}) = \frac{A}{\left(1 - \xi_1^2/v_1^2 - \xi_2^2/v_2^2\right)^{1/2}} \tag{3.4.26}$$

Substituting this expression into (3.4.17) and carrying out the integrations, we get

$$F_{\mathrm{P}}(t, \mathbf{m}) = \frac{2\pi A v_1 v_2 t^2}{\left(1 - v_1^2 m_1^2/v_{\mathrm{P}}^2 - v_2^2 m_2^2/v_{\mathrm{P}}^2\right)^2}$$

$$\tag{3.4.27}$$

$$F_{\mathrm{S}}(t, \mathbf{m}) = \frac{2\pi A v_1 v_2 t^2}{\left(1 - v_1^2 m_1^2/v_{\mathrm{S}}^2 - v_2^2 m_2^2/v_{\mathrm{S}}^2\right)^2}$$

Let $v_1 > v_2$. Then v_1 is the largest fracture velocity. It is seen from (3.4.27) that $F_{\mathrm{P}}(t, \mathbf{m})$ and $F_{\mathrm{S}}(t, \mathbf{m})$ are maximum at $m_1 = 1$, $m_2 = m_3 = 0$, that is, when the ray direction coincides with the direction of maximum fracture velocity.

Expressions (3.4.17) and (3.4.27) depend on time only through factor t^2. This means that the record of displacements for body waves near the first arrivals should have a parabolic shape and that of velocities should be straight lines. This conclusion is in drastic contradiction to the actual shapes of seismograms, which are irregular oscillatory curves. This contradiction is due to the assumptions that the medium and stress drop are homogeneous and that the fracture velocity is independent of time. The latter is incompatible with any fracture criterion. The first two assumptions hold only if the microstructure of the medium is neglected, that is, for a smooth phenomenological description. In terms of Section 3.3, it can be said that (3.4.17) and subsequent expressions apply only to the coherent part of the radiation. If the microinhomogeneity of the medium is taken into account, the fracture propagation and slip will be discrete and discontinuous. The pulses close to the initiation time (which corresponds to the first arrivals) due to individual discrete fractures must be few and not overlap in time. The number of pulses increases as the fracture grows (due to the increase in fracture area and the length of its edge); their duration also increases (smoother fracture propagation), and their interference becomes more important. This is manifested in an increase of apparent period on the seismogram. More important reasons for the oscillatory appearance of the seismogram are the dispersion and scattering during propagation and filtering by the instrument.

In a smooth description of fracture, the slip rate $\dot{a}(\mathbf{x}, t)$ has its maximum value (infinite) near the edge. It can be expected that the pulses radiated during discrete elementary fractures and associated with discontinuous variations in the slip velocity will have the maximum amplitude near the fracture edge. Then the pattern of the radiation field will again be as shown in Figure 3.5, where the individual wavelets now represent the waves from each individual discrete jump of fracture. In this case, however, while examining the incoherent part of radiation, one should assume that the individual pulses do not overlap in time, so that the density of fronts is related not to the magnitude of displacement but to the visible period of the record (to the sequence of pulses), which is maximum in the direction of maximum fracture velocity.

Thus, there is one more possibility of determining the orientation of the fault plane based on the properties of the first arrivals of body waves (now separately, for P and S waves).

To do this, it is necessary to

1. reduce the observations to those for a homogeneous medium and a standard distance from the source;
2. find the fault-plane solution from the signs of first arrivals (position of nodal planes) and calculate the theoretical amplitude at every station from (3.4.5);

3. exclude the directivity for a concentrated source by dividing the records by the theoretical amplitudes; and
4. select the stations where the relative amplitude and frequency obtained are maximum. Then the nodal plane, that closest to the ray corresponding to the station, will be the actual fault plane and the projection of this ray on the fault plane will give the direction of maximum fracture propagation velocity.

The quantitative theory of incoherent radiation, the formulation of which would require the application of statistical methods, is beyond the scope of this book. However, it is seen from the above discussion that, without such a theory, it is impossible to gain sufficient insight into the short-wave source radiation and, particularly, the structure of first arrivals. Unfortunately, short-wave radiation is influenced not only by the microinhomogeneity of the fracture propagation process but also by the microstructure of the medium during wave propagation. Therefore, it is not yet clear if this method has practical application.

3.5 Spectral representation of the inverse problem

In this section, we will approach the solution of equations (3.3.21) and (3.3.29) assuming that the fracture surface is a plane whose orientation is known beforehand by some independent means. Let us reproduce equations (3.3.21) and (3.3.29) for convenience:

$$m_l m_m \int_{\Sigma_1} \dot{m}_{lm}\left(\mathbf{x}, t + \frac{m_i x_i}{v_P}\right) dS = 4\pi \rho v_P^3 A^P(t, \mathbf{m}) \tag{3.3.21}$$

$$(m_k m_l - \delta_{kl}) m_m \int_{\Sigma_1} \dot{m}_{lm}\left(\mathbf{x}, t + \frac{m_i x_i}{v_S}\right) dS = 4\pi \rho v_S^3 A_k^S(t, \mathbf{m}) \tag{3.3.29}$$

Here, $A^P(t, \mathbf{m})$ and $A_k^S(t, \mathbf{m})$ are the records of longitudinal and transverse waves at a station corresponding to the ray direction $\mathbf{m}$ at the focus, corrected for geometric spreading and attenuation. Assume that these functions are known for all directions $\mathbf{m}$. This is, of course, a very idealized assumption, but here we are interested only in the solvability of the system of equations (3.3.21) and (3.3.29).

Let us take the Fourier transform with respect to time of equations (3.3.21) and (3.3.29), that is, their spectral representation. The spectrum of $f(t)$ will be denoted by $f(\omega)$, the difference being explicitly indicated by the argument t or ω. For example,

$$A^P(\omega, \mathbf{m}) = \int_0^\infty e^{i\omega t} A^P(t, \mathbf{m}) \, dt \tag{3.5.1}$$

so that $A^P(\omega, \mathbf{m})$ is the spectrum of the P wave corresponding to the direction $\mathbf{m}$. Applying (3.5.1) to equations (3.3.21) and (3.3.29) we obtain

$$m_l m_m \int_{\Sigma_1} m_{lm}(\mathbf{x}, \omega) e^{-i\omega m_i x_i / v_P} \, dS = \frac{4\pi i v_P^3}{\omega} \rho A^P(\omega, \mathbf{m}) \tag{3.5.2}$$

and

$$(m_k m_l - \delta_{kl}) m_m \int_{\Sigma_1} m_{lm}(\mathbf{x}, \omega) e^{-i\omega m_i x_i / v_S} \, dS = \frac{4\pi i v_S^3}{\omega} \rho A_k^S(\omega, \mathbf{m}) \tag{3.5.3}$$

Let us change the variables in equation (3.5.2) by

$$\xi = \frac{\omega \mathbf{m}^{\parallel}}{v_P} = (\mathbf{m} - (\mathbf{mn})\mathbf{n}) k_P \tag{3.5.4}$$

where $k_P = \omega / v_P$ is the wave number of the P wave and $\mathbf{n}$, as before, is the unit normal to the fault plane. Vector $\mathbf{m}$ can easily be expressed in terms of ξ. For this, remember that $\mathbf{m}$ is a unit vector, that is,

$$(\mathbf{m}^{\parallel})^2 + (\mathbf{mn})^2 = 1$$

whence

$$(\mathbf{mn}) = \pm \left(1 - (\mathbf{m}^{\parallel})^2\right)^{1/2} = \pm \left(1 - k_P^2 \xi^2\right)^{1/2}$$

where the sign is to be chosen according to which side of the fault plane the vector $\mathbf{m}$ lies, and $\xi = \sqrt{\xi_i \xi_i}$ is the modulus of vector ξ. Now

$$\mathbf{m} = \mathbf{m}^{\parallel} + \mathbf{n}(\mathbf{mn}) = \frac{1}{k_P} \left(\xi \pm \mathbf{n}(k_P^2 - \xi^2)^{1/2}\right) \tag{3.5.5}$$

In equation (3.5.3), let us put

$$\xi = (\mathbf{m} - (\mathbf{mn})\mathbf{n}) k_S \tag{3.5.6}$$

$$\mathbf{m} = \frac{1}{k_S} \left(\xi + \mathbf{n}(k_S^2 - \xi^2)^{1/2}\right) \tag{3.5.7}$$

where $k_S = \omega/v_S$. Equations (3.5.2) and (3.5.3) now take the form

$$\left(\xi_l \pm n_l\left(k_P^2 - \xi^2\right)^{1/2}\right)\left(\xi_m \pm n_m\left(k_P^2 - \xi^2\right)^{1/2}\right)\int_{\Sigma_1} m_{lm}(\mathbf{x}, \omega)e^{-i\xi_i x_i}\,dS$$

$$= 4\pi i\rho v_P\omega A^P\left(\omega, \frac{1}{k_P}\left(\xi \pm \mathbf{n}\left(k_P^2 - \xi^2\right)^{1/2}\right)\right)$$

$$\left[\left(\xi_k \pm n_k\left(k_S^2 - \xi^2\right)^{1/2}\right)\left(\xi_l \pm n_l\left(k_S^2 - \xi^2\right)^{1/2}\right) - k_S^2\delta_{kl}\right] \tag{3.5.8}$$

$$\times\left(\xi_m \pm n_m\left(k_S^2 - \xi^2\right)^{1/2}\right)\int_{\Sigma_1} m_{lm}(\mathbf{x}, \omega)e^{-i\xi_i x_i}\,dS$$

$$= 4\pi i\rho\omega^2 A_k^S\left(\omega, \frac{1}{k_S}\left(\xi \pm \mathbf{n}\left(k_S^2 - \xi^2\right)^{1/2}\right)\right)$$

Substituting (3.2.17),

$$m_{lm}(\mathbf{x}, \omega) = \lambda\delta_{lm}a_i(\mathbf{x}, \omega)n_i + \mu\left(a_l(\mathbf{x}, \omega)n_m + a_m(\mathbf{x}, \omega)n_l\right)$$

into (3.5.8), we get, after some simple manipulations,

$$\left[\left(k_S^2 - 2\xi^2\right)n_l \pm 2\xi_l\left(k_P^2 - \xi^2\right)^{1/2}\right]\int_{\Sigma_1} a_l(\mathbf{x}, \omega)e^{-i\xi_i x_i}\,dS$$

$$= \frac{4\pi i v_P\omega}{v_S^2}A^P\left(\omega, \frac{1}{k_P}\left(\xi \pm \mathbf{n}\left(k_P^2 - \xi^2\right)^{1/2}\right)\right)$$

$$\left[\left(n_k\left(k_S^2 - 2\xi^2\right) \pm 2\xi_k\left(k_S^2 - \xi^2\right)^{1/2}\right)\left(\xi_l \pm n_l\left(k_S^2 - \xi^2\right)^{1/2}\right)\right. \tag{3.5.9}$$

$$\left.\mp k_S^2\left(k_S^2 - \xi^2\right)^{1/2}\delta_{kl}\right]\int_{\Sigma_1} a_l(\mathbf{x}, \omega)e^{-i\xi_i x_i}\,dS$$

$$= \frac{4\pi i\omega^2}{v_S^2}A_k^S\left(\omega, \frac{1}{k_S}\left(\xi \pm \mathbf{n}\left(k_S^2 - \xi^2\right)^{1/2}\right)\right)$$

Multiplying the second of these equations by n_k, we get

$$\left[\xi_l\left(k_S^2 - 2\xi^2\right) \mp 2\xi^2\left(k_S^2 - \xi^2\right)^{1/2}n_l\right]\int_{\Sigma_1} a_l(\mathbf{x}, \omega)e^{-i\xi_i x_i}\,dS$$

$$= \frac{4\pi i\omega^2}{v_S^2}n_k A_k^S\left(\omega, \frac{1}{k_S}\left(\xi \pm \mathbf{n}\left(k_S^2 - \xi^2\right)^{1/2}\right)\right) \tag{3.5.10}$$

Due to the condition $m_k A_k^S = 0$, multiplication of this equation by ξ_k gives the same result (3.5.10). Finally, multiplying the second equation of (3.5.9) by the vector $e_{krs} n_r \xi_s$ (where e_{krs} is the unit antisymmetric tensor), we get

$$
e_{lrs} n_r \xi_s \int_{\Sigma_1} a_l(\mathbf{x}, \omega) e^{-i\xi_i x_i} \, dS = \mp \frac{4\pi i e_{klm}}{\left(k_S^2 - \xi^2\right)^{1/2}} n_l \xi_m
$$
$$
\times A_k^S \left(\omega, \frac{1}{k_S}\left(\xi \pm \mathbf{n}\left(k_S^2 - \xi^2 \right)^{1/2} \right) \right)
$$

(3.5.11)

The first equation of (3.5.9) and equation (3.5.10) contain the scalar products $n_l a_l$ and $\xi_l a_l$. Let us transform these equations into a pair of equations, each of which contains only one of these products. To do this, we multiply the first equation of (3.5.9) by $(k_S^2 - 2\xi^2)$ and equation (3.5.10) by $\pm 2(k_P^2 - \xi^2)^{1/2}$ and eliminate the term proportional to ξ_l from the equations obtained. Then we get

$$
\int_{\Sigma_1} n_l a_l(\mathbf{x}, \omega) e^{-i\xi_i x_i} \, dS
$$
$$
= \frac{4\pi i k_S^2}{R(\omega, \xi)} \left[\frac{1}{k_P}\left(k_S^2 - 2\xi^2 \right) A^P \mp 2\left(k_P^2 - \xi^2 \right)^{1/2} n_k A_k^S \right]
$$
(3.5.12)

where

$$
R(\omega, \xi) = \left(k_S^2 - 2\xi^2 \right)^2 + 4\xi^2\left(k_P^2 - \xi^2 \right)^{1/2}\left(k_S^2 - \xi^2 \right)^{1/2}
$$
(3.5.13)

For simplicity, the arguments of the functions A^P and A_k^S are omitted.

Next, multiplying the first equation of (3.5.9) by $\pm 2\xi^2(k_S^2 - \xi^2)^{1/2}$ and equation (3.5.10) by $k_S^2 - 2\xi^2$ and adding the equations so obtained, we get

$$
\int_{\Sigma_1} \xi_l a_l(\mathbf{x}, \omega) e^{-i\xi_i x_i} \, dS
$$
$$
= \frac{4\pi i k_S^2}{R(\omega, \xi)} \left[\pm \frac{2}{k_P}\xi^2\left(k_S^2 - \xi^2 \right)^{1/2} A^P + \left(k_S^2 - 2\xi^2 \right) n_k A_k^S \right]
$$

(3.5.14)

We now take into account that the vectors n_k, ξ_l, and $e_{krs} n_r \xi_s$ are orthogonal. Therefore, the vector $a_l(\mathbf{x}, \omega)$ can be resolved along these three vectors:

$$a_k(\mathbf{x}, \omega) = n_k(n_l a_l) + \frac{1}{\xi^2}\xi_k(\xi_l a_l) + \frac{1}{\xi^2}e_{krs}n_r\xi_s(e_{lpq}n_p\xi_q a_l)$$

(3.5.15)

Multiplying equation (3.5.11) by $e_{krs}n_r\xi_s/\xi^2$, equation (3.5.12) by n_k, and equation (3.5.14) by ξ_k/ξ^2, and using equation (3.5.15), we get the vector equation

$$\int_{\Sigma_1} a_k(\mathbf{x}, \omega)e^{-i\xi_i x_i}\, dS = F_k(\omega, \xi)$$

(3.5.16)

where

$$F_k(\omega, \xi) = 4\pi i \left\{ n_k k_S^2 \frac{(k_S^2 - 2\xi^2)A^P \mp 2k_P(k_P^2 - \xi^2)^{1/2}n_l A_l^S}{k_P R(\omega, \xi)} \right.$$

$$+ \xi_k k_S^2 \frac{k_P(k_S^2 - 2\xi^2)n_l A_l^S \mp 2\xi^2(k_S^2 - \xi^2)^{1/2}A^P}{\xi^2 k_P R(\omega, \xi)}$$

$$\left. \mp \frac{e_{krs}n_r\xi_s e_{lpq}n_p\xi_q A_l^S}{\xi^2(k_S^2 - \xi^2)^{1/2}} \right\}$$

(3.5.17)

Equation (3.5.16) is valid only for those values of ξ and ω for which functions $A^P(\omega, \mathbf{m})$ and $A_k^S(\omega, \mathbf{m})$ are determined, that is, within the cone

$$\xi \le k_P < k_S$$

(3.5.18)

For these values of ξ, the vector $F_k(\omega, \xi)$ is calculated from the spectra of body waves. Integration in equation (3.5.16) is carried out over the fracture area Σ_1 at the instant when the earthquake stops, outside which $a_k \equiv 0$. The size and shape of the area Σ_1 are not known beforehand and must be found as part of the solution. Therefore, it is convenient to extend the integration over the entire fault plane Σ, that is,

$$\int_{\Sigma} a_k(\mathbf{x}, \omega)e^{-i\xi_i x_i}\, dS = F_k(\omega, \xi)$$

(3.5.19)

under the additional condition that the solution $a_k(\mathbf{x}, \omega)$ is sought in the class of functions having finite support with respect to $\mathbf{x}$.

If the condition of the absence of opening, that is, $a_k n_k = 0$, is used, then expression (3.5.17) for the right side of equation (3.5.16) can be simplified. It follows from this condition that $F_k(\omega, \xi)n_k = 0$. Multiplying (3.5.17) by n_k, we get

$$\left(k_S^2 - 2\xi^2\right)A^P \mp 2k_P\left(k_P^2 - \xi^2\right)^{1/2}A_l^S n_l = 0 \tag{3.5.20}$$

From this equation, we express A^P in terms of $A_l^S n_l$, and substitute it to (3.5.17) to get

$$F_k(\omega, \xi) = 4\pi i\left\{\xi_k k_S^2 \frac{n_l}{\xi^2\left(k_S^2 - 2\xi^2\right)} \pm \frac{e_{krs}e_{lpq}n_r n_p \xi_s \xi_q}{\xi^2\left(k_S^2 - \xi^2\right)^{1/2}}\right\}A_l^S \tag{3.5.17a}$$

This expression has two advantages. First, it is known from the observation of S waves only, and in its evaluation the combination of the values of A^P and A_k^S at identical values of ξ, that is, at different values of **m**, is not required. Because of the discreteness of any seismic network, stations having conjugate records of P and S waves would be difficult to find. Second, for S waves the values of ξ are not limited to the range of (3.5.19) but to a wider range $\xi < k_S$, which is also important.

From the theoretical point of view, however, these improvements are of little importance, and in what follows we will assume that the right side of equation (3.5.16) is given in the domain (3.5.18) and refrain from using the hypothesis on the absence of opening.

3.6　Solution to the basic equation; uniqueness theorem

We shall fix the system of coordinates in such a way that axes x_1 and x_2 lie in the fault plane and axis x_3 is oriented along its normal. Then, it is clear that in equation (3.5.19), $x_3 = \xi_3 = 0$, and this equation can be rewritten as

$$\iint_{-\infty}^{\infty} a_k(x_1, x_2, \omega)e^{-i(x_1\xi_1 + x_2\xi_2)}\,dx_1\,dx_2 = F_k(\omega, \xi_1, \xi_2) \tag{3.6.1}$$

Here the left side represents a double Fourier transform of the vector $a_k(\mathbf{x}, \omega)$ by spatial coordinates, or a triple Fourier transform of the vector $a_k(\mathbf{x}, t)$ by all three variables (triple spectrum of the displacement jump vector). The right side can be calculated from the spectra of the body waves for every ray by the formula (3.5.17). Consequently, from seismic observations one can calculate the triple spectrum of the displacement jump vector. If this could be done for all real values of ξ_1 and ξ_2, then equation (3.6.1) could be solved by taking the inverse Fourier transform:

$$a_k(\mathbf{x}, \omega) = \frac{1}{4\pi^2} \int\!\!\!\int\limits_{-\infty}^{\infty} F_k(\omega, \boldsymbol{\xi}) e^{i\xi_i x_i} \, d\xi_1 \, d\xi_2 \qquad (3.6.2)$$

However, $F_k(\omega, \boldsymbol{\xi})$ is known from observations only in the domain (3.5.18), that is, for

$$\xi_1^2 + \xi_2^2 \leq \omega^2/v_P^2 \qquad (3.6.3)$$

Thus, the problem is reduced to the continuation of the vector $F_k(\omega, \boldsymbol{\xi})$ from the circle (3.6.3) onto the infinite plane $\xi_1\xi_2$, for each value of frequency ω. As already stated, the unknown vector $a_k(\mathbf{x}, \omega)$ must vanish outside a region Σ_1 in the plane $x_1 x_2$ (outside the fracture area). Furthermore, in Chapter 2, the existence of the energy integral was assumed for the solution $u_i(\mathbf{x}, t)$. This assumption implies that $a_k \in L_2$ space, and therefore $a_k(\mathbf{x}, \omega)$ must also belong to L_2. Thus, the solution of equation (3.6.1) is required to be a vector, the components of which have finite support and belong to the L_2 space on the plane $x_1 x_2$, which imposes rigid constraints on the properties of the right side of equation (3.6.1). These constraints are established by the following theorem of Plancherel and Polya (see Titchmarsh, 1948), which is stated here without proof:

Let the function $f(\xi_1, \xi_2)$ be an element of L_2 space within the whole plane $(\xi_1\xi_2)$. In order that its Fourier transform

$$\tilde{f}(x_1, x_2) = \frac{1}{4\pi^2} \int\!\!\!\int\limits_{-\infty}^{\infty} f(\xi_1, \xi_2) e^{i(\xi_1 x_1 + \xi_2 x_2)} \, d\xi_1 \, d\xi_2 \qquad (3.6.4)$$

vanish almost everywhere outside a finite region Σ of the plane $x_1 x_2$, it is necessary and sufficient that $f(\xi_1, \xi_2)$ be continuable into the whole space of the complex variables $\zeta_1 = \xi_1 + i\eta_1$ and $\zeta_2 = \xi_2 + i\eta_2$ as an entire function of finite order. At the same time, the smallest convex region $\tilde{\Sigma}$, containing Σ, is determined by the condition

$$k(\phi) = h(\phi) \qquad (3.6.5)$$

where $k(\phi)$ is the support function of the area $\tilde{\Sigma}$, and $h(\phi)$ is the P-indicator function of the entire function $f(\zeta_1, \zeta_2)$. The P-indicator function of the entire function $f(\zeta_1, \zeta_2)$ is defined as the limit

$$h(\phi) = \lim_{\substack{R \to \infty \\ \xi_1^2 + \xi_2^2 < R^2}} \sup \frac{\ln f(\xi_1 + iR\cos\phi, \xi_2 + iR\sin\phi)}{R} \qquad (3.6.6)$$

and the support function $k(\phi)$ of the convex region $\tilde{\Sigma}$ as

$$k(\phi) = \sup_{x_1, x_2 \in \tilde{\Sigma}} (x_1 \cos \phi + x_2 \sin \phi) \qquad (3.6.7)$$

In order to apply this theorem, it is also necessary to refer to the Plancherel theorem for Fourier integrals, which establishes, in particular, that the Fourier transform of a function of L_2 space is also an element of L_2. Then it follows from (3.6.1) that $F_k(\omega, \xi) \in L_2$, together with $a_k(\mathbf{x}, \omega)$, and we may apply the Plancherel–Polya theorem to equation (3.6.2) with the result that $F_k(\omega, \zeta_1, \zeta_2)$ is continuable as an entire function $F_k(\omega, \zeta_1, \zeta_2)$ of the two complex variables ζ_1 and ζ_2. The P-indicator of this function coincides with the support function of the smallest convex region $\tilde{\Sigma}$, containing the fracture area Σ_1. For the present purposes, it will be sufficient to know that $F_k(\omega, \zeta_1, \zeta_2)$ is an entire function with respect to each of its arguments, ζ_1 and ζ_2.

By Cauchy's theorem the analytic continuation is unique and can be constructed as follows. Let (ξ_1^0, ξ_2^0) be some point within the circle (3.6.3). Then $F_k(\omega, \zeta_1, \zeta_2)$ is given as the double Taylor series:

$$F_k(\omega, \zeta_1, \zeta_2)$$
$$= \sum_{m=0}^{\infty} \sum_{n=0}^{\infty} \frac{1}{n!} \frac{1}{m!} \frac{\partial^{(n+m)}}{\partial \xi_1^n \partial \xi_2^m} F_k(\omega, \xi_1^0, \xi_2^0)(\zeta_1 - \xi_1^0)^n (\zeta_2 - \xi_2^0)^m$$

$$(3.6.8)$$

Because $F_k(\omega, \zeta_1, \zeta_2)$ is an entire function, the series (3.6.8) converges for all complex values ζ_1 and ζ_2. Consequently, for the continuation of $F_k(\omega, \xi_1, \xi_2)$, it is necessary and sufficient to know this function precisely in any neighborhood of any arbitrary point (ξ_1^0, ξ_2^0) of the circle (3.6.3), and it is all the more sufficient to know it on the entire circle. In particular, constraining the variables ζ_1 and ζ_2 to be real valued, we obtain the continuation of $F_k(\omega, \xi_1, \xi_2)$ onto the infinite plane (ξ_1, ξ_2), and this continuation, as stated earlier, is unique. Substituting this continuation $F_k(\omega, \xi_1, \xi_2)$ into equation (3.6.2), we obtain the solution $a_k(\mathbf{x}, \omega)$ of the equation (3.6.1), and applying the inverse Fourier transform over time,

$$a_k(\mathbf{x}, t) = \frac{1}{2\pi} \int_{-\infty}^{\infty} e^{-i\omega t} a_k(\mathbf{x}, \omega) \, d\omega \qquad (3.6.9)$$

we find the distribution of the displacement jump across the fracture as a function of coordinates and time.

Let us apply a Fourier transform over time to equation (3.6.2):

$$a_k(\mathbf{x}, t) = \frac{1}{4\pi^2} \iint\limits_{-\infty}^{\infty} F_k(t, \xi) e^{-i\xi_i x_i} \, d\xi_1 \, d\xi_2 \qquad (3.6.10)$$

The function $a_k(\mathbf{x}, t)$ also belongs to L_2 and vanishes outside the fracture area $\Sigma(t)$, contained within Σ. Consequently, the Plancherel–Polya theorem may also be applied to equation (3.6.10); that is, the function $F_k(t, \xi_1, \xi_2)$ is continuable for every t to an entire function of finite order. Let the P-indicator function of F_k be $h(\phi, t)$. Then, by the Plancherel–Polya theorem, it coincides with the support function of the (convex) region $\tilde{\Sigma}(t)$, that is, $k(\phi, t) = h(\phi, t)$. The equation for the boundary of the region $\tilde{\Sigma}$ can be obtained from the support function $h(\phi, t)$ in the following manner. We use polar coordinates on the plane $x_1 x_2$:

$$x_1 = r \cos \phi; \qquad x_2 = r \sin \phi \qquad (3.6.11)$$

Then from definition (3.6.7) of the support function, it follows that the straight line

$$r \cos(\phi - \phi_0) = h(\phi_0, t) \qquad (3.6.12)$$

is tangent to the curve $L(t)$, the boundary of $\tilde{\Sigma}(t)$ [under the condition, of course, that the curve $L(t)$ is smooth]. Consequently, $L(t)$ is the envelope of the family of straight lines (3.6.12). The equation of the envelope is obtained if ϕ_0 is eliminated from equation (3.6.12) and from its derivative with respect to ϕ_0,

$$r \sin(\phi - \phi_0) = h'_{\phi_0}(\phi_0, t) \qquad (3.6.13)$$

The parametric equation of the envelope is obtained if, in equations (3.6.12) and (3.6.13), we return to Cartesian coordinates and solve the resulting system with respect to x_1 and x_2:

$$\begin{aligned}
x_1 &= \cos \phi_0 h(\phi_0, t) - \sin \phi_0 h'_{\phi_0}(\phi_0, t) \\
x_2 &= \sin \phi_0 h(\phi_0, t) + \cos \phi_0 h'_{\phi_0}(\phi_0, t)
\end{aligned} \qquad (3.6.14)$$

where $0 \le \phi_0 \le 2\pi$.

Let us formulate the principal results of this section in the form of a theorem:

Theorem 1.   The solution $a_k(\mathbf{x}, t)$ of the inverse problem of earthquake source theory, belonging to the class L_2, exists when and only when the components of vector $F_k(\omega, \xi)$, determined by equation (3.5.17), are continuable as the entire function $F_k(\omega, \zeta_1, \zeta_2)$ of finite order of the two complex variables ζ_1, ζ_2. Under this condition, the solution is unique and is given by the expression (3.6.10), where

$$F_k(t, \xi) = \frac{1}{2\pi} \int_{-\infty}^{\infty} e^{-i\omega t} F_k(\omega, \xi) \, d\omega \qquad (3.6.15)$$

and $F_k(t, \xi)$ is also continuable for any $t > 0$ to an entire vector function. In particular, the equations of motion of the boundary of the convex envelope of the fracture area are expressed in terms of the P-indicator function of $F_k(t, \xi)$ by (3.6.14). The slip functions $a_k(\mathbf{x}, t)$ of the form derived from (3.4.18) and (3.4.25) belong to the class L_2.

3.7 Instability of the solution

The fact that $F_k(\omega, \xi)$ belongs to the space of entire functions, as established in the preceding section, applies to ideally exact observations, which in addition must be available for every direction of the ray **m**. In reality, the values of functions $A^P(\omega, \mathbf{m})$ and $A_k^S(\omega, \mathbf{m})$ are measured at a finite number of points and with errors. Therefore, the functions $F_k(\omega, \xi)$ calculated from observations will not, generally speaking, be entire, and they cannot be used as the right side of equation (3.6.1), because the conditions for the existence of the solution may be violated. We denote the functions obtained from observations by $F_k^*(\omega, \xi)$. Then we are faced with the problem of approximating the vector function $F_k^*(\omega, \xi)$ by an entire function of the finite order $F_k(\omega, \xi) \in L_2$, for real values of ξ_1, ξ_2. Without loss of generality, it may be assumed that $F_k^*(\omega, \xi)$ is a continuous vector function of ξ_1 and ξ_2. We will proceed as follows. First, we will construct the class of entire functions of simple structure, everywhere dense within the set of all functions of L_2 on $\xi_1 \xi_2$, continuable as entire functions of finite order for all complex values of these variables. Then we will show that the class constructed is also everywhere dense within the set of all continuous functions given in the finite domain (3.6.3). Thereby, the possibility of approximating the continuous function F^*, in the domain of its definition, by an entire function possessing the necessary properties will be established.

The Plancherel–Polya theorem states that every function $f(\xi_1, \xi_2) \in L_2$ on the plane $\xi_1 \xi_2$ and continuable as an entire function of finite order is a Fourier transform of a function $\tilde{f}(x_1, x_2)$ with finite support also belonging to L_2. It is always possible to find a square $(-a < x_1 < a; -a < x_2 < a)$ containing the support of the function $\tilde{f}(x_1, x_2)$. Let us select the smallest of these. Obviously the continuation of $\tilde{f}(x_1, x_2)$ to this square (zero outside its support) also belongs to L_2. Therefore, $\tilde{f}(x_1, x_2)$ can be expanded on this square into a convergent double Fourier series:

$$\tilde{f}(x_1, x_2) = \sum_{m,\,n=-\infty}^{\infty} c_{mn} \exp[\pi(imx_1 + inx_2)/a] \qquad (3.7.1)$$

This means that for any $\varepsilon > 0$, there exist integers M and N such that

$$\int\!\!\int_{-a}^{a} \left| \tilde{f}(x_1, x_2) - \sum_{m=-M}^{M} \sum_{n=-N}^{N} c_{mn} \exp[\pi i(mx_1 + nx_2)/a] \right|^2 dx_1\, dx_2 < \varepsilon$$

$$(3.7.2)$$

The function $f(\xi_1, \xi_2)$ can be expressed as the Fourier transform of $\tilde{f}(x_1, x_2)$, that is, as the inverse of (3.6.4):

$$f(\xi_1, \xi_2) = \int\!\!\int_{-a}^{a} \tilde{f}(x_1, x_2) \exp[-i(\xi_1 x_1 + \xi_2 x_2)]\, dx_1\, dx_2 \qquad (3.7.3)$$

The corresponding Fourier transform of the trigonometric polynomial, (3.6.2), has the form

$$\sum_{m=-M}^{M} \sum_{n=-M}^{N} c_{mn} \int\!\!\int_{-a}^{a} \exp\left\{ i\left[\left(\pi\frac{m}{a} - \xi_1\right) x_1 + \left(\pi\frac{n}{b} - \xi_2\right) x_2 \right] \right\} dx_1\, dx_2$$

$$= \sum_{m=M}^{M} \sum_{n=-N}^{N} a^2 c_{mn} \frac{4\sin(a\xi_1 - \pi m)\sin(a\xi_2 - \pi n)}{(a\xi_1 - \pi m)(a\xi_2 - \pi n)} \qquad (3.7.4)$$

Thus, the function

$$\sum_{m=-M}^{M} \sum_{n=-N}^{N} c_{mn} \exp[\pi i(mx_1 + nx_2)/a]$$

is the Fourier transform of the function

$$\sum_{m=-M}^{M} \sum_{n=-N}^{N} 4a^2 c_{mn} \frac{\sin(a\xi_1 - \pi m)\sin(a\xi_2 - \pi n)}{(a\xi_1 - \pi m)(a\xi_2 - \pi n)}$$

Applying the Plancherel theorem to this pair of the functions and using the inequality (3.7.2), we get

$$\iint\limits_{-\infty}^{\infty} \left| f(\xi_1, \xi_2) - \sum_{m=-M}^{M} \sum_{n=-N}^{N} 4a^2 c_{mn} \right.$$

$$\left. \times \frac{\sin(a\xi_1 - m\pi)\sin(a\xi_2 - n\pi)}{(a\xi_1 - m\pi)(a\xi_2 - n\pi)} \right|^2 d\xi_1 \, d\xi_2 < \varepsilon \qquad (3.7.5)$$

for any ε, for sufficiently large values of M and N. Thus, any function $f(\xi_1, \xi_2) \in L_2$ and continuable to an entire function of finite order can be approximated, to any precision, by the finite sum

$$f_{MN}^a(\xi_1, \xi_2) = \sum_{m=-M}^{M} \sum_{n=-N}^{N} 4a^2 c_{mn} \frac{\sin(a\xi_1 - m\pi)\sin(a\xi_2 - n\pi)}{(a\xi_1 - m\pi)(a\xi_2 - n\pi)}$$

Reducing the terms of this sum to a common denominator, we represent f_{MN}^a in the form

$$f_{MN}^a(\xi_1, \xi_2) = P_{2M2N}(\xi_1, \xi_2) \Psi_{MN}^a(\xi_1, \xi_2) \qquad (3.7.6)$$

where $P_{2M2N}(\xi_1, \xi_2)$ is a polynomial of degree $(2M + 2N)$, and

$$\Psi_{MN}^a(\xi_1, \xi_2) = \frac{\sin a\xi_1 \sin a\xi_2}{a^2 \xi_1 \xi_2 \Pi_{m=0}^{M} \Pi_{n=0}^{N} \left(1 - a^2\xi_1^2/(\pi^2 n^2)\right)\left(1 - a^2\xi_2^2/(\pi^2 m^2)\right)} \qquad (3.7.7)$$

Remembering the infinite product expansion of the function $\sin(ax)$, we conclude that

$$\lim_{\substack{M \to \infty \\ N \to \infty}} \Psi_{MN}^a(\xi_1, \xi_2) = 1 \qquad (3.7.8)$$

and that the convergence to this limit is uniform in any finite domain of variation of ξ_1, ξ_2.

We now consider an arbitrary function $\phi(\xi_1, \xi_2)$, continuous within the circle, $\xi_1^2 + \xi_2^2 < c^2$. Using the Weirstrass theorem on polynomial approximation and the property (3.7.8) of the function Ψ_{MN}^a, it is easy to show that, for any positive ε and fixed a, such M and N can always be found that

$$\sup_{\xi_1^2 + \xi_2^2 < c^2} |\phi(\xi_1, \xi_2) - f_{MN}^a(\xi_1, \xi_2)| < \varepsilon$$

In other words, the set of functions $f_{MN}^a(\xi_1, \xi_2)$ for any fixed value of a is everywhere dense in the space of function continuous on any circle $\xi_1^2 + \xi_2^2 < c^2$.

We shall apply this result to the function $F_k^*(\omega, \xi)$, which is known and continuous within the circle (3.7.3). As was just demonstrated, it can be uniformly approximated, to any precision, by a function from the set (3.7.6):

$$F_k^*(\omega, \xi) = \Phi_{kMN}(\omega, \xi)\Psi_{MN}^a \equiv F_{kMN}^a(\omega, \xi) \qquad (3.7.9)$$

where a is an arbitrary real number.

The functions $F_{MN}^a(\omega, \xi)$ are entire functions of finite order, the P indicator function of these functions being $h_{MN}(\phi) = (|\cos \phi| + |\sin \phi|)a$; that is, the support of the Fourier transform of $F_{MN}^a(\omega, \xi)$ is contained in a square of side a. Substituting $F_{MN}^a(\omega, \xi)$ into the right side of equation (3.7.1) in place of $F_k(\omega, \xi)$, we arrive at the following theorem.

Theorem 2. For any positive number a, a distribution of $a_k(\mathbf{x}, \omega)$ can always be found that is concentrated within a square of any side a and reproduces the observed far-field body wave radiation to any given accuracy; this approximation is uniform with respect to the coordinates of the observation points and within any finite frequency range.

Physically this means that it is possible to construct a model of an earthquake source of arbitrarily small size, for which the far-field body wave radiation is indistinguishable from the observed. On the basis of equation (3.6.1), it is impossible to determine (due to unavoidable errors of observation) even the source size. In other words, the solution of the inverse problem of earthquake source theory is unstable. It is possible to overcome this instability only by applying to the solution $a_k(\mathbf{x}, t)$ some set of more rigid constraints than belonging to L_2. It can be shown that, within a specified precision, it is always possible to choose $F_{kMN}(\omega, \xi_1, \xi_2)$ such that, for sufficiently large M and N, its norm in L_2 exceeds any given number. This means that it is possible to construct two distributions of $a_k(\mathbf{x}, t)$, producing indistinguishable far-field body wave radiation and, at the same time, having arbitrarily different L_2 norms.

The instability of the inverse problem of source theory is not unexpected when we take into account the physical meaning of the vector ξ. From equation (3.5.4), it is seen that ξ is the projection of the wave number of the P wave $\mathbf{k}_P = \mathbf{m}\omega/v_P = 2\pi\mathbf{m}/\lambda_P$ on the fault plane. Therefore, the magnitude of the vector ξ does not exceed the value $2\pi/\lambda_P$. On the other hand, $1/\xi$ is the wavelength of spatial harmonics into which $a_k(\mathbf{x}, \omega)$ may be expanded. Thus, the impossibility of continuing $F_k(\omega, \xi)$ beyond the circle (3.6.3) implies simply that from observations it is not possible to obtain the details of the distribution of

$a_k(\mathbf{x}, \omega)$ on a spatial scale, which is less than the minimum observed wavelength.

The instability of the inverse problem implies two conclusions that are important for the study of the earthquake source. The first of these is that it is impossible to solve the far-field inverse problem by the method of trial and error, that is, by defining a class of simple source models and choosing from it a model that agrees best with observations, because it is possible to arrive at conclusions about the motions at the source that are arbitrarily different from reality. Yet there is no way of overcoming this instability other than by constraining the class of models. However, in choosing such a class of models, we must be *a priori* certain that they contain all physically realizable functions $a_k(\mathbf{x}, t)$. Thus, the constraint of the class of source models must be based on some physical information independent of seismic observations.

Such information can be provided only by consideration of the process as a physical one, that is, in the framework of fracture mechanics. Fracture mechanics implies, in particular, that the displacement jump vector cannot be constant over the whole fracture area, because this would violate the condition of finiteness of the elastic energy. In other words, the class of models in the form of Volterra dislocations does not even intersect the class of physically permissible models. This, however, does not mean that physically permissible distributions of displacement jump that are almost constant within the fracture area do not exist, but in every case, for the time being, there is no physical basis for expecting that such a distribution can be found. Currently, the only implication of fracture mechanics that could be applied to source theory is the limit on the fracture propagation velocity. For investigating the solvability of the inverse problem the spectral representation was not only convenient but unavoidable. In earthquake source theory, however, we are concerned with the radiation of seismic waves within a broad frequency range so that the wave vector $\boldsymbol{\xi}$ belonging to the visibility range (3.6.3) for one frequency will be far beyond it for another lower frequency, and for very low frequencies (tending toward zero), the visibility domain is reduced to a single point $\boldsymbol{\xi} = 0$. The main disadvantage of investigating the inverse problem for every frequency separately is that it becomes impossible to take into account even the constraint on fracture velocity.

The second conclusion follows directly from the first. Namely, if the spectral representation does not permit the determination of even the dimensions of the fracture area because of the instability of the inverse problem, the method of fault-plane determination from stopping phases becomes even more important. Spectral representation smears out these secondary arrivals (since spectra are always calculated for a limited range of frequencies). In every case, seismograms in the time domain contain more information about the source than do their spectra.

The instability of the inverse problem, which was established by means of the Fourier transform method, obviously does not depend on this method. Why, then, does the stopping-phase method provide the ability to estimate the source dimensions and the fracture velocity whereas the solution of the inverse problem does not? There are two reasons. The first of these is that, in formulating the inverse problem, we were restricted to observations of seismic waves in the far field; that is, we neglected the ratio of the source size to the hypocentral distance as well as the differences in the directions of the waves arriving at a given observation point from various parts of the source. The second is that determinations of arrivals on seismograms correspond to measurements at very high frequencies of the order of the sampling frequency. With an increase in frequency, the visibility range $\xi < |\omega/v_P|$ expands, which results in a reduction of the instability of the solution. Let us show that the far-field approximation does not affect the possibility of application of the stopping-phase method. Let some arrival at N stations be identified as a stopping phase, that is, as a short pulse radiated at a point $\mathbf{x}_0$ of the fracture at an instant t_0. Let the arrival times of this phase, measured from the instant of first arrival, be $t_{(\kappa)}$, and the directions of rays corresponding to each station be $\mathbf{m}_{(\kappa)}$ ($\kappa = 1, \ldots, N$). Then, in the far-field approximation we will have, for P waves,

$$t_\kappa + \frac{\mathbf{m}_{(\kappa)}\mathbf{x}_0}{v_P} = t_0, \qquad \kappa = 1, \ldots, N \tag{3.7.10}$$

that is, N equations to solve for the four unknowns – the three components of vector $\mathbf{x}_0$ and time t_0. Consequently, the minimum number of stations needed is four, and if the orientation of the fault plane is known, then even three stations are sufficient. So if the arrivals are identified as stopping phases, the instability of the problem in the far field disappears. It is obvious that here regularization is just the procedure of identification of phases, in particular the distinguishing of useful phases from the noise (reflections, microseisms, etc.). This remark applies to the many examples of this general approach reported in the literature (Kanamori and Stewart, 1978; House and Boatwright, 1980; Kikuchi and Kanamori, 1982; Ruff, 1983; Hartzell and Heaton, 1983, 1986; and others).

As an example of the instability of the inverse problem, let us consider the 1979 Imperial Valley, California, earthquake, the source characteristics of which were determined by various authors using different methods. Most of these attempts were essentially trial-and-error fitting of the observed seismograms. However, Olson and Apsel (1982) attempted a formal inversion using the relation (3.2.11); the left side of the equation was obtained from observations, the Green functions were calculated for a layered half-space, and the fault slip was inverted by a least squares method. The results of this formal inversion were very different from the solutions obtained by the trial-and-error method (e.g., Hartzell and Heaton, 1983), even though the same seismograms were used!

3.8   Smoothed solution to the inverse problem

At present the possibility of formulating sufficiently rigid restrictions on slip distribution and history at the source so as to regularize the inverse problem seems unlikely. The general principles of fracture mechanics alone are insufficient for this purpose simply due to their generality, and there are no data available to make them more specific. At the same time, however, rigid constraints chosen *a priori* may distort the solution unacceptably.

When one is solving typical unstable problems, such as the analytic continuation of potential fields, the additional constraint consists essentially of the requirement that the solution be sufficiently smooth. For potential fields, this requirement is natural. In contrast, one cannot expect the slip distribution to be smooth, neither in space nor in time. The situation seems to be hopeless. There exists, however, a way out, and it does not even involve a physical hypothesis. Namely, let us resign ourselves to being unable to determine the actual slip distribution and instead try to determine a smoothed slip distribution. We will select a smoothing kernel $K(\mathbf{x}, t)$, say, and convolve it with the unknown slip distribution $a_i(\mathbf{x}, t)$. Then

$$\bar{a}_i(\mathbf{x}, t) = \int_0^t \int_\Sigma a_i(\mathbf{x}, \tau) K(\mathbf{x} - \mathbf{y}, t - \tau) \, dS \, d\tau$$

where Σ is the whole fault plane and $\bar{a}_i$ the smoothed slip. Now if we could reformulate the inverse problem in terms of $\bar{a}_i$, we would obtain the additional restriction requiring the solution to be smooth and would apply standard methods of solving unstable problems. Unfortunately, to do so we would have to integrate the seismograms by ray direction, which, with a realistic number of stations, would be rather impracticable. To overcome this difficulty, let us suppose that smoothing the slip distribution over time only will produce a smooth function of space coordinates as well. This seems plausible since during the dynamic process the irregularities in sliding and fracturing must propagate. Accordingly, we define the smoothed slip as

$$\bar{a}_i(\mathbf{x}, t) = \int_0^t K(t - \tau) a_i(\mathbf{x}, t)\, d\tau \tag{3.8.1}$$

Now let us rewrite equation (3.5.19) in the time domain,

$$\int_\Sigma a_k\left(\mathbf{x}, t + \frac{\mathbf{mx}}{v_\mathrm{S}}\right) = F_k(t, \mathbf{m}) \tag{3.8.2}$$

where the opening displacement is assumed to vanish and $F_k(t, \mathbf{m})$ is obtained from (3.5.17a) as

$$F_k(t, \mathbf{m}) = 4\pi v_\mathrm{S}\left(\frac{-m_k n_l}{1 - 2(m_j n_j)^2} \pm \frac{e_{krs} e_{lpq} n_r n_p m_s m_q}{m_j n_j}\right) \int_0^t A_l^S(t, \mathbf{m})\, dt \tag{3.8.3}$$

and is simply proportional to the seismogram of the S wave at the station corresponding to the ray direction $\mathbf{m}$, integrated by time. Now convolving both sides of (3.8.2) with $K(t)$, we obtain

$$\int_\Sigma \int_0^t a_k(\mathbf{x}, \tau) K\left(t + \frac{\mathbf{mx}}{v_\mathrm{S}} - \tau\right) d\tau\, dS$$

$$= \int_0^t F_k(\tau, \mathbf{m}) K(\tau, \mathbf{m}) K(t - \tau)\, d\tau$$

The inner integral is equal to $\bar{a}_k(\mathbf{x}, t + \mathbf{mx}/v_\mathrm{S})$, and we finally get

$$\int_\Sigma \bar{a}_k\left(\mathbf{x}, t + \frac{\mathbf{mx}}{v_\mathrm{S}}\right) dS = \bar{F}_k(t, \mathbf{m}) \tag{3.8.4}$$

where the right side is the S-wave seismogram filtered by the kernel

$$K_1(t) = \int_0^t K(\tau)\, d\tau$$

and is known from observations. The equation (3.8.4) has the same form as the original equation (3.8.1), but the unknown function is now smooth, which allows it to be determined numerically.

Another possible approach consists of determining spatial moments of the slip distribution a_k instead of the distribution itself. This approach was essentially suggested by Backus and Mulcahy (1976a, b). To analyze this possibility, we start with the spectral representation of the basic equation (3.6.1). Let us develop both sides of it into power series in terms of ξ_1 and ξ_2 in the vicinity of the point $\xi = 0$ and equate terms of the same order from both sides. On the right side we will have the series (3.6.8), where $\xi_1^0 = \xi_2^0 = 0$, and on the left side Taylor's series for the exponent will be substituted. Then we will obtain

$$\iint\limits_{-\infty}^{\infty} a_k(x_1, x_2, \omega) x_1^{s_1} x_2^{s_2}\, dx_1\, dx_2 = (i)^{s_1+s_2} \frac{\partial^{s_1+s_2}}{\partial \xi_1^{s_1}\, \partial \xi_2^{s_2}} F_k(\omega, \xi_1, \xi_2)\Bigg|_{\substack{\xi_1=0 \\ \xi_2=0}} \tag{3.8.5}$$

Here the left side represents the spatial moment of a_k of the order $s_1 s_2$, where s_1 and s_2 are arbitrary positive integers. The variables ξ_1 and ξ_2 are related to the ray direction **m** by equations (3.5.4) and (3.5.6) for P and S waves, respectively. Let us represent them in the form

$$\xi_1 = \omega p_1; \qquad \xi_2 = \omega p_2 \tag{3.8.6}$$

where

$$p_1 = \frac{m_1}{v_P}; \qquad p_2 = \frac{m_2}{v_P} \tag{3.8.7}$$

for P waves and

$$p_1 = \frac{m_1}{v_S}; \qquad p_2 = \frac{m_2}{v_S} \tag{3.8.8}$$

for S waves.

The functions F_k expressed in terms of ω, p_1, and p_2 represent combinations of P- and S-wave seismograms after exclusion of the point source radiation patterns. Their dependence on p_1 and p_2 reflects the deviation of the source radiation pattern from that of a concentrated dipole.

We denote the moments on the left side of (3.8.5) by $a_k^{s_1 s_2}(\omega)$. Then this equation can be rewritten as

$$a_k^{s_1 s_2}(\omega) = \left(\frac{i}{\omega}\right)^{s_1+s_2} \frac{\partial^{s_1+s_2}}{\partial p_1^{s_1}\, \partial p_2^{s_2}} F_k(\omega, p_1, p_2)\Bigg|_{\substack{p_1=0 \\ p_2=0}} \tag{3.8.9}$$

or, returning to the time domain,

$$a_k^{s_1 s_2}(t) = \frac{\partial^{s_1+s_2}}{\partial p_1^{s_1}\, \partial p_2^{s_2}} \int_0^t F_k(\tau, p_1, p_2)(t-\tau)^{s_1+s_2-1}\, d\tau\Bigg|_{\substack{p_1=0 \\ p_2=0}} \tag{3.8.10}$$

In these equations, the instability is not wholly eliminated, since one must numerically differentiate the observed wave field by ray direction. This is, however, a standard problem and can be solved, say, by approximating the observed field by spherical harmonics and then differentiating the resulting approximation. Evaluation of higher-order moments is difficult not only because of the necessary differentiation but also because of the suppression of high frequencies by the factor $\omega^{-s_1-s_2}$ in equation (3.8.9). Obviously, the lower the frequency, the less the radiation pattern will deviate from that of the point source. The zero-order moment does not require differentiation by ray direction or integration of seismograms, so its determination is most practicable. In fact, it differs only in notation from the seismic moment tensor, which has become a routinely determined earthquake source parameter. The theory of the seismic moment tensor is considered in detail in the next chapter.

4

SEISMIC MOMENT TENSOR

4.1 Long-wave approximation

It was shown in Section 3.2 that, with respect to the radiation of seismic waves, each element of fracture at the earthquake source is equivalent to an elastic dipole having a moment tensor $m_{lm}\,dS\,dt$, where the density m_{lm} of the dipole moment is equal to

$$m_{lm}(\mathbf{x}, t) = \lambda \delta_{lm} a_i(\mathbf{x}, t) n_i(\mathbf{x})$$
$$+ \mu \big(a_l(\mathbf{x}, t) n_m(\mathbf{x}) + a_m(\mathbf{x}, t) n_l(\mathbf{x}) \big) \qquad (4.1.1)$$

Fracture as a whole is, generally speaking, not equivalent to a dipole source if the wavelength is comparable to the fracture dimension or less than it.

We write the spectra of body wave displacement using the relations (3.3.33) through (3.3.36) as

$$u^{\mathrm{P}}(\mathbf{x}_1, \omega) = \frac{m_l m_m e^{i\omega T_{\mathrm{P}}(\mathbf{x}_1, 0)} \omega}{4\pi i \rho v_{\mathrm{P}}^3 R_{\mathrm{P}}(\Delta, h)} \int_{\Sigma} m_{lm}(\mathbf{x}, \omega) e^{-i k_{\mathrm{P}} x_i m_i}\, dS$$

$$u^{\mathrm{SV}}(\mathbf{x}_1, \omega) = \frac{(m_k m_l - \delta_{kl}) z_k m_m \omega e^{i\omega T_{\mathrm{S}}(\mathbf{x}_1, 0)}}{4\pi i \rho v_{\mathrm{S}}^3 R_{\mathrm{SV}}(\Delta, h) \big(1 - m_r^2 z_r^2 \big)^{1/2}}$$

$$\times \int_{\Sigma} m_{lm}(\mathbf{x}, \omega) e^{-i k_{\mathrm{S}} x_i m_i}\, dS \qquad (4.1.2)$$

$$u^{\mathrm{SH}}(\mathbf{x}_1, \omega) = \frac{e_{lij} m_i z_j m_m \omega e^{i\omega T_{\mathrm{S}}(\mathbf{x}_1, 0)}}{4\pi i \rho v_{\mathrm{S}}^3 R_{\mathrm{SH}}(\Delta, h) \big(1 - m_r^2 z_r^2 \big)^{1/2}}$$

$$\times \int_{\Sigma} m_{lm}(\mathbf{x}, \omega) e^{i k_{\mathrm{S}} x_i m_i}\, dS$$

These relations are alternative forms of equations (3.5.2) and (3.5.3). Radiation from a concentrated dipole source would be obtained if we put $m_{lm} = M_{lm}(\omega)$ in equation (4.1.2), where $M_{lm}(\omega)$ is the spectrum of the concentrated dipole moment. In other words, to obtain the radiation from a concentrated dipole it is necessary to replace the integrals in equations (4.1.2) by the tensor $M_{lm}(\omega)$, which is independent of the direction of the ray.

Let d be the maximum size of fracture area Σ_1. Then it is obvious that

$$|k_P x_i m_i| < \frac{2\pi d}{\lambda_P}; \qquad |k_S x_i m_i| < \frac{2\pi d}{\lambda_S} \qquad (4.1.3)$$

where λ_P and λ_S are the P and S wavelengths for frequency ω. For long waves, when

$$k_P d = \frac{2\pi d}{\lambda_P} \ll 1 \quad \text{and} \quad k_S d = \frac{2\pi d}{\lambda_S} \ll 1 \qquad (4.1.4)$$

we have

$$e^{-ik_P x_i m_i} = 1 + O(k_P d)$$
$$e^{-ik_S x_i m_i} = 1 + O(k_S d) \qquad (4.1.5)$$

and expression (4.1.2) coincides with the expressions for radiation from a point dipole source with the spectrum of moment tensor

$$M_{lm}(\omega) = \int_{\Sigma_1} m_{lm}(\mathbf{x}, \omega)\, dS \qquad (4.1.6)$$

Thus, for body waves whose wavelengths are many times larger than the size of the fracture at the source, it is equivalent to the concentrated dipole source.

Using (4.1.1) we obtain

$$M_{lm}(\omega) = \lambda \delta_{lm} \int_{\Sigma_1} n_i(\mathbf{x}) a_i(\mathbf{x}, \omega)\, dS$$
$$+ \mu \int_{\Sigma_1} (n_l(\mathbf{x}) a_m(\mathbf{x}, \omega) + n_m(\mathbf{x}) a_l(\mathbf{x}, \omega))\, dS \qquad (4.1.7)$$

The conditions (4.1.4) under which the source can be replaced by a point source, can be rewritten as

$$T \gg \frac{2\pi}{v_P} \quad \text{and} \quad T \gg \frac{2\pi}{v_S} \qquad (4.1.8)$$

where T is the wave period. If, as expected in most cases, the fracture velocity v is not much less than the limiting velocity and, hence, not much less than v_P and v_S, the following condition would be fulfilled together with condition (4.1.8):

$$T \gg 2\pi t_m \qquad (4.1.9)$$

where t_m is the maximum fracture duration at the source.

In general, the condition (4.1.9) is more rigid that (4.1.4) or (4.1.8). With this condition, it is possible to obtain more specific conclusions on the dependence of $M_{lm}(\omega)$ on frequency. In fact,

$$a_i(\mathbf{x}, \omega) = \int_0^\infty a_i(\mathbf{x}, t) e^{i\omega t}\, dt = \int_0^{t_m} a_i(\mathbf{x}, t) e^{i\omega t}\, dt + a_i^1(\mathbf{x}) \int_{t_m}^\infty e^{i\omega t}\, dt$$

where $a_i^1(\mathbf{x})$ is the final displacement jump after the earthquake, that is, at $t > t_m$. Then under condition (4.1.9) we get

$$a_i(\mathbf{x}, \omega) = \frac{i}{\omega} e^{i\omega t_m} a_i^1(\mathbf{x}) + O(1) \qquad (4.1.10)$$

or, with the same accuracy,

$$a_i(\mathbf{x}, \omega) = \frac{i}{\omega} a_i^1(\mathbf{x}) + O(1) \qquad (4.1.11)$$

Substituting this expression in (4.1.7), we get

$$M_{lm}(\omega) \approx \frac{i}{\omega} M_{0lm} \qquad \text{for} \quad \omega t_m \ll 1 \qquad (4.1.12)$$

where

$$M_{0lm} = \lambda \delta_{lm} \int_{\Sigma_1} a_i^1(\mathbf{x}) n_i(\mathbf{x})\, dS + \mu \int_{\Sigma_1} \left(a_l^1(\mathbf{x}) n_m(\mathbf{x}) + a_m^1(\mathbf{x}) n_l(\mathbf{x}) \right)\, dS$$
$$(4.1.13)$$

Substituting (4.1.12) into (4.1.2), we get

$$e^{i\omega T_P(\mathbf{x}_1,0)} u^P(\mathbf{x}_1, \omega) \approx \frac{m_l m_m}{4\pi\rho v_S^3 R_P(\Delta, h)} M_{0lm}$$

$$e^{-i\omega T_S(\mathbf{x}_1,0)} u^{SV}(\mathbf{x}_1, \omega) \approx \frac{(m_k m_l - \delta_{kl}) z_k m_m}{4\pi\rho v_S^3 R_{SV}(\Delta, h)\left(1 - m_r^2 z_r^2\right)^{1/2}} M_{0lm} \quad (4.1.14)$$

$$e^{-i\omega T_S(\mathbf{x}_1,0)} u^{SH}(\mathbf{x}_1, \omega) \approx \frac{e_{lij} m_i z_j m_m}{4\pi\rho v_S^3 R_{SH}(\delta, h)\left(1 - m_r^2 z_r^2\right)^{1/2}} M_{0lm}$$

Thus, the displacement spectra of seismic body waves are flat at low frequencies. As $\omega \to 0$, they tend toward constant values, which are determined by the tensor M_{0lm} only. The tensor M_{0lm} is called the seismic moment tensor (Randall, 1971). Expression (4.1.13) for the seismic moment tensor was obtained without making the assumption of vanishing of the opening displacement and for an arbitrary, not necessarily planar, fracture surface Σ_1.

The trace of M_{0lm} is equal to

$$M_{0ii} = (3\lambda + 2\mu) \int_{\Sigma_1} a_i^1 n_i \, dS = 3k \, \Delta V^1 \tag{4.1.15}$$

where k is the bulk modulus and ΔV^1 is the final change in volume due to opening displacement. For pure shear, $\Delta V^1 = 0$ and M_{0lm} is a deviator. Moreover, if the fracture area is planar, then n does not depend on $\mathbf{x}$ and expression (4.1.13) takes the form

$$M_{0lm} = \mu S(n_l \bar{a}_m + m_m \bar{a}_l) \tag{4.1.16}$$

where S is the area of Σ_1, and $\bar{a}_i$ is the average slip vector

$$\bar{a}_i = \frac{1}{S} \int_{\Sigma_1} a_i^1(\mathbf{x}) \, dS \tag{4.1.17}$$

Denote the length of vector $\bar{a}$ by a, and the unit vector having the same direction by b_i

$$\bar{a}_i = ab_i; \qquad b_i b_i = 1 \tag{4.1.18}$$

Then M_{0lm} can be written as

$$M_{0lm} = M_0(n_l b_m + n_m b_l) \tag{4.1.19}$$

where

$$M_0 = \mu a S \tag{4.1.20}$$

Here M_0 is the seismic moment of the earthquake defined by Aki (1966). It can be seen from (4.1.14) and (4.1.20) that u^P vanishes if the ray $\mathbf{m}$ is perpendicular to either of the vectors $\mathbf{n}$ or $\mathbf{b}$; that is, these vectors are normal to the P-wave nodal planes at low frequencies. Therefore, the tensor M_{0lm} can be calculated if the seismic moment M_0 and the fault-plane solution for long waves are known. In determining the seismic moment M_0, the comparison of theoretical and observed spectral amplitudes at low frequencies is commonly used. In calculating theoretical amplitudes the fault-plane solution (vectors $\mathbf{n}$ and $\mathbf{b}$) for the first arrivals is used. A drawback of this approach is that the fault-plane solution from first arrivals gives the directions of the slip and the normal to the fault surface in the neighborhood of the point where fracture was initiated and for times close to the initiation time. However, vectors $\mathbf{n}$

and **b**, defined by (4.1.19), are the directions of the normal and the final slip vector, respectively, averaged over the entire fracture area. The initial slip might occur in another direction or along some plane other than the main fracture. The likelihood of such deviation is indicated by the discrepancy between fault-plane solutions from the signs of first arrivals at regional and teleseismic distances that is sometimes observed. Therefore, the direct determination of the seismic moment tensor M_{0lm} is more reliable than the determination of seismic moment M_0 using fault-plane solutions from first arrivals.

4.2 Determination of the seismic moment tensor from observations

The formulas for the long-wavelength approximation for body waves only were written because the corresponding expressions for surface waves are rather involved. It can be shown, however, that in the case of surface waves the source, in the long-wavelength approximation, is equivalent to the dipole source of the same type. That is, the radiation from the source is the same as the radiation produced by the distribution of body forces,

$$\rho f_i(\mathbf{x}, t) = M_{0lm} \frac{\partial}{\partial x_l} \delta_{im} \delta(\mathbf{x}) H(t) \qquad (4.2.1)$$

where $H(t)$ is the Heaviside function, provided that the wavelength is many times larger than the fracture dimension and the period is much longer than the source duration. In fact, (4.1.2) and (4.1.14) are also not of practical use because the filtering effects of the medium and the instrument have been omitted. The practical approach to the calculation of the seismic moment tensor must be based on the generation of numerical synthetic seismograms for realistic models of velocity structure and attenuation within the earth. A detailed description of such a procedure using normal modes is given in the review of Dziewonski and Woodhouse (1983).

Using the symmetry of the tensor M_{0lm}, rewrite (4.2.1) as

$$\rho f_i(\mathbf{x}, t) = M_{0lm} \rho f_{ilm}(\mathbf{x}, t) \qquad (4.2.2)$$

where

$$\rho f_{ilm}(\mathbf{x}, t) = \frac{1}{2} \left(\delta_{im} \frac{\partial}{\partial x_l} \delta(\mathbf{x}) + \delta_{il} \frac{\partial}{\partial x_m} \delta(\mathbf{x}) \right) H(t) \qquad (4.2.3)$$

Thus, the equivalent dipole source is a linear combination of six independent balanced dipoles (4.2.3).

In equations (4.2.1) through (4.2.3), the origin of coordinates $x = 0$
coincides with the hypocenter. We choose the x_1 axis vertically upward,
the x_2 axis to the north, and the x_3 axis to the east from the hypocenter.

Let $A_{lm}^{Q,n}(\omega)$ be the displacement spectrum for a wave of the type Q
(where Q = P, SV, SH, R, L, etc.), calculated for each of the elementary
dipoles (4.2.3) and for each of the stations $n = 1, \ldots, N$ at which
observations are available. As $\omega \to 0$, each of the functions of $A_{lm}^{Q,n}$ must
tend toward the constant value $A_{0lm}^{Q,n}$. Because linear elasticity theory is
used, the observed amplitude $A_0^{Q,n}$ at each station for low frequencies
should be, according to (4.2.2), a linear combination of amplitudes $A_{0lm}^{Q,n}$:

$$A_{0lm}^{Q,n} M_{0lm} = A_0^{Q,n} \qquad (4.2.4)$$

In these equations the coefficients $A_{0lm}^{Q,n}$ are known from theoretical
calculation and $A_0^{Q,n}$ are known from observations. Thus, the relations
(4.2.4) comprise a system of simultaneous linear equations for the six
components of the seismic moment tensor M_{0lm}. The number of equations is equal to the product of the number of independent waves
(subscript Q) and the number of stations at which records of these waves
are available in the low-frequency range.

If it is assumed that the normal to the fracture area and the slip
direction are constant, it is possible to use the vectors **n** and **b** from the
fault-plane solution. Then, substituting expression (4.1.19) into (4.2.4),
we get the equation

$$A_{0lm}^{Q,n}(n_l b_m + n_m b_l) M_0 = A_0^{Q,n} \qquad (4.2.5)$$

which contains only one unknown quantity, the seismic moment M_0.

Let the seismic moment tensor for an earthquake be determined using
(4.2.4). Then, by calculating its trace

$$M_{0ii} = 3k\,\Delta V$$

it is possible to verify whether there is a significant amount of opening
displacement, that is, the hypothesis as to the absence of opening. If
there is no opening displacement, the tensor M_{0lm} is the deviator. Let
the principal axes of M_{0lm} be represented by three unit vectors l_{ik},
$k = 1, 2, 3$. If the fracture area is planar, (4.1.19) implies that the
principal axes are related to the vectors **n** and **b** by

$$l_{i1} = \frac{n_i + b_i}{\sqrt{2}}; \qquad l_{i2} = \frac{n_i - b_i}{\sqrt{2}}; \qquad l_{i3} = e_{ikl} n_k b_l \qquad (4.2.6)$$

and the principal values are equal to $M_{01} = M_0$, $M_{02} = -M_0$, $M_{03} = 0$. That is, for a plane fracture surface, the mean principal value of the deviator of tensor M_{0lm} vanishes. This mean principal value provides a check on whether the fracture area is really planar.

4.3   Seismic moment versus stress estimates at the source

One of the fundamental applications of the concept of seismic moment is the supposed possibility of estimating the average value of stresses acting at the earthquake source. Two estimates of the stress conditions at the source were used: stress drop $\Delta\sigma$, which is equal to the difference of initial (σ^0) and final (σ^1) stress values on the fault,

$$\Delta\sigma = \sigma^0 - \sigma^1 \tag{4.3.1}$$

and effective stress σ, which is equal to half the sum of the initial and final stresses,

$$\sigma = \frac{\sigma^0 + \sigma^1}{2} \tag{4.3.2}$$

Of course, here, σ^0 and σ^1 mean some average values over the fracture area. The relation between stress drop and seismic moment M_0 can easily be obtained from dimensional analysis based on the expression (4.1.20) for M_0. In fact, it follows from the linearity of the basic equations that $\Delta\sigma$ and a must be proportional to one another. Thus, we have the following relation,

$$a = c\frac{\Delta\sigma}{\mu}l \tag{4.3.3}$$

where l is some dimension of the fracture area and c is a dimensionless constant depending on the shape of the fracture area at the source. This constant has been estimated (Chinnery, 1969; Aki, 1972a; Kanamori and Anderson, 1975; Brune, Archuleta, and Hartzell, 1979; Das, 1981) to be between 0.2 and 0.4 for the most important cases. Furthermore, l is proportional to the square root of the fracture area: $l \approx S^{1/2}$. Thus, from (4.1.19) and (4.3.3) we obtain, with sufficient accuracy for estimation purposes

$$M_0 = c\,\Delta\sigma\,S^{3/2} \tag{4.3.4}$$

or

$$\Delta\sigma = \frac{M_0}{c}S^{-3/2} \tag{4.3.5}$$

Thus, the seismic moment can be used to estimate the stress drop at the source.

If one assumes that the fault plane coincides with the plane of initial maximum shear stress, the principal axes of the seismic moment tensor (4.2.12) would coincide with the principal axes of initial stress σ_{ij}, and under the additional assumption of total stress drop ($\sigma^1 = 0$), the average initial stress tensor can be written as

$$\sigma_{ij}^0 = \frac{1}{c} S^{-3/2} M_{0ij} \tag{4.3.6}$$

This relation raises many doubts. In fact, fractures at earthquake sources occur along preexisting faults. Even if their orientation coincided in the distant past with the plane of maximum shear stress, it is quite likely that this coincidence is not preserved at present, because the orientation of stress changes with time, even if only because of earthquakes themselves, not to mention the variations in the tectonic environment. Moreover, for shear fracture the stress drops only partially due to friction, and only on the fracture surface, and the final stress σ_{ij} is not coaxial with σ_{ij}^0. Consequently, one may speak only of the stress drop vector and not about a tensor. Therefore, the principal axes of the moment tensor (fault-plane solution) do not coincide with the principal axes of any stress tensor in the vicinity of the source. These remarks are essentially a repetition of the reasoning in Section 3.2 for the moment density tensor m_{lm} and its relation to the initial stress tensor σ_{ij}^0. It is important to note here that confusing the nodal planes of P waves with the planes of maximum initial shear stress obscures the relation of the fault-plane solution to the seismic moment tensor, which, as will be seen in Section 4.5, has a profound geophysical meaning, being a measure of earthquake contribution to tectonic deformation.

The definition of effective stress σ originated from the following formula for the seismic energy,

$$E_q = \eta \sigma a S = \eta \sigma \frac{M_0}{\mu} \tag{4.3.7}$$

where E_q is the seismic energy of the earthquake and η the so-called seismic efficiency. This equation defines essentially the quantity

$$\eta \sigma = \frac{\mu E_q}{M_0} \tag{4.3.8}$$

which is called the "apparent stress" (Wyss and Brune, 1968, 1971). If it were possible to determine the seismic efficiency η independently, the effective stress σ could be found from the values of energy and moment by (4.3.8), and using the stress drop $\Delta\sigma$, defined by (4.3.5), it would be possible to calculate the initial stress σ^0 and final stress σ^1 at the source. However, from the linearity of the basic equations it follows that displacements during an earthquake, measured from the initial state of equilibrium, cannot be dependent on the initial and final stress states separately, but only on their difference τ_{ij} (see Chapter 1). This means that σ^1 or σ^0 separately or σ, which is their arithmetic mean, cannot, in principle, be determined from seismic observations. Therefore, the possibility of determining σ from formula (4.3.7) is only illusory. For a systematic comparison of some of these estimates of stress release for a particular earthquake, one may refer to Boatwright (1984).

Equation (4.3.7) was obtained by identifying the change in the potential energy of the earth due to an earthquake with the sum of seismic energy and frictional losses along the fault, assuming that these losses could be accounted for by the seismic efficiency η. This implies that seismic energy is not so simply related to the potential energy change as has sometimes been assumed. In the following section, we will discuss in detail the concept of seismic energy, which will enable us to clarify the meaning of "apparent stress" $\eta\sigma$.

4.4 Seismic energy

Seismic energy is defined as the the total energy transmitted by seismic waves through a surface S_0 surrounding the source, that is,

$$E_q = -\int_0^{t_m} dt \int_{S_0} \tau_{ij}\dot{u}_i n_j \, dS \qquad (4.4.1)$$

where t_m is the source duration and the waves reflected from the free surface of the earth are neglected, which is equivalent to replacing the earth by an infinite elastic medium. In practice, this formula is specified by choosing a sphere of sufficiently large radius and by approximating the seismogram by a set of sinusoids. Thus, Galitzin's formula is obtained (Bullen, 1963). Equation (4.4.1) is the definition of seismic energy as a physical quantity, since it describes the method of measuring it.

In (4.4.1), τ_{ij} is the stress calculated from the seismic wave displacement using Hooke's law:

$$\tau_{ij} = c_{ijkl}u_{k,l}$$

That is, it is the difference between the current (σ_{ij}) and initial (σ_{ij}^0) stress tensors.

First we evaluate the change in the potential energy in a volume V bounded by S_0, due to an earthquake. The final elastic energy density is

$$w^1 = w_0 + \sigma_{ij}^0 u_{i,j}^1 + \tfrac{1}{2}c_{ijkl}u_{i,j}^1 u_{k,l}^1 = w_0 + \sigma_{ij}^0 u_{i,j}^1 + \tfrac{1}{2}\tau_{ij}^1 u_{i,j}^1$$

$$= w_0 + \tfrac{1}{2}\left(\sigma_{ij}^1 + \sigma_{ij}^0\right)u_{i,j}^1$$

where σ_{ij}^1 is the final stress tensor and w_0 the initial potential energy density. The potential energy change consists of the changes in elastic and gravitational energies during an earthquake,

$$\Delta E_{\mathrm{p}} = \int_V \left(w^1 - w_0 - \rho f_i u_i^1\right) dV$$

where f_i is the gravitational acceleration and u_i^1 the final displacement. Substituting the expression for $(w^1 - w^0)$ into the previous formula, we get

$$\Delta E_{\mathrm{p}} = \int_V \left\{\tfrac{1}{2}\left(\sigma_{ij}^1 + \sigma_{ij}^0\right)u_{i,j}^1 - \rho f_i u_i^1\right\} dV$$

Since the medium is initially and finally at rest under the action of the same body force ρf_i, we have

$$\tfrac{1}{2}\left(\sigma_{ij}^1 + \sigma_{ij}^0\right)_{,j} = -\rho f_i$$

and the expression for ΔE_{p} can be rewritten as

$$\Delta E_{\mathrm{p}} = \frac{1}{2}\int_V \left\{\left(\sigma_{ij}^1 + \sigma_{ij}^0\right)u_i^1\right\}_{,j} dV \tag{4.4.2}$$

Integrating by parts, we get

$$\Delta E_{\mathrm{p}} = \frac{1}{2}\int_{S_0}\left(\sigma_{ij}^1 + \sigma_{ij}^0\right)u_i^1 n_j\, dS - \frac{1}{2}\int_{\Sigma_1}\left(\sigma_{ij}^1 + \sigma_{ij}^0\right)a_i^1 n_j\, dS \tag{4.4.3}$$

As the radius of S_0 tends toward infinity, the first term in (4.4.3) tends toward zero and, for sufficiently large radius of S_0, may be neglected. Then

$$\Delta E_{\mathrm{p}} = -\frac{1}{2}\int_{\Sigma_1}\left(\sigma_{ij}^1 + \sigma_{ij}^0\right)a_i n_j\, dS \tag{4.4.4}$$

If it is assumed that the tractions $\sigma_{ij}^1 n_j$ and $\sigma_{ij}^0 n_j$ are collinear with a_i

and independent of the position on Σ_1, we would get

$$\Delta E_p = -\sigma a S \tag{4.4.5}$$

and if it is assumed further that the seismic energy E_q is equal to the decrease in the potential energy of the earth $-\Delta E_p$ multiplied by seismic efficiency η, we would get (4.3.7). This assumption, however, has no relation to the definition (4.4.1) of the seismic energy as a physical quantity. However, if (4.3.7) is considered to be the definition of seismic efficiency, it is of course true. But then it would be necessary to obtain a separate expression for this quantity to make (4.3.7) meaningful. Usually it is assumed that the decrease in potential energy E_p is equal to the sum of the seismic energy E_q and work of frictional forces A_f on the fault,

$$E_q + \Delta E_p + A_f = 0 \tag{4.4.6}$$

where

$$A_f = \int_{\Sigma_1} \sigma_{(f)i} a_i \, dS \tag{4.4.7}$$

This expression implies that neither the magnitude $\sigma_{(f)}$ of frictional force nor its orientation depends on time during an earthquake. Furthermore, considering $\sigma_{(f)}$ to be constant all over the fracture surface, we get:

$$A_f = \sigma_{(f)} a S \tag{4.4.8}$$

and equations (4.4.6) and (4.4.5) will give

$$E_q = (\sigma - \sigma_{(f)}) a \tag{4.4.9}$$

Then it is obvious that

$$\eta = 1 - \frac{\sigma_{(f)}}{\sigma} \tag{4.4.10}$$

and the problem is reduced to the determination of the frictional stress $\sigma_{(f)}$.

However, equation (4.4.6) is essentially a new definition of seismic energy E_q. Before using the relations (4.4.9) and (4.4.10), which follow from this definition, it is necessary to verify whether definitions (4.4.1) and (4.4.6) are equivalent. At first glance, (4.4.6) appears to be a trivial consequence of the energy conservation law. Actually this is not so. Even if it were assumed that frictional work is the only loss of energy during an earthquake, the energy conservation law would be

$$\Delta E_{S_0} + \Delta E_p + A_f = 0 \tag{4.4.6a}$$

where ΔE_{S_0} is the energy released from the volume V through its surface

S_0 during an earthquake and is equal to

$$\Delta E_{S_0} = -\int_0^{t_m} dt \int_{S_0} \sigma_{ij} n_j \dot{u}_i \, dS \qquad (4.4.11)$$

This differs from expression (4.4.1) for E_q in that instead of stress perturbation τ_{ij} it contains the total stress σ_{ij}. Their difference is

$$\Delta E_{S_0} - E_q = \int_0^{t_m} dt \int_{S_0} \sigma_{ij}^0 n_j \dot{u}_i \, dS = \int_{S_0} \sigma_{ij}^0 u_i^1 \, dS \qquad (4.4.12)$$

which represents the work of initial stresses (over the surface S_0) on the displacement u_i^1 due to an earthquake. Generally speaking, this value is nonzero because, as is known, large internal stresses are present practically everywhere in the earth. Therefore, generally speaking, (4.4.6) does not agree with the definition of seismic energy (4.4.1). Even if (4.4.12) could be neglected, equation (4.4.6a) does not account for energy losses during fracture.

To obtain the relation of seismic energy E_q to quantities pertaining to the source, let us first of all consider the rate of change of elastic energy in the volume V during an earthquake:

$$\dot{W} = \frac{d}{dt} \int_V \frac{1}{2} \left(\sigma_{ij} + \sigma_{ij}^0 \right) u_{i,j} \, dV \qquad (4.4.13)$$

Because the volume V contains the propagating fracture edge where the stress and particle velocity have a singularity of order $\frac{1}{2}$, it cannot be differentiated with respect to time under the integral sign in (4.4.13). Therefore, we proceed as in Section 2.2: We exclude the toroidal volume V_ε surrounding the fracture edge and write

$$\dot{W} = \lim_{\varepsilon \to 0} \frac{d}{dt} \int_{V - V_\varepsilon} \frac{1}{2} \left(\sigma_{ij} + \sigma_{ij}^0 \right) u_{i,j} \, dV$$

Now we can differentiate under the sign of integration:

$$\dot{W} = \lim_{\varepsilon \to 0} \int_{V - V_\varepsilon} \left(\frac{1}{2} \dot{\sigma}_{ij} u_{i,j} + \frac{1}{2} \left(\sigma_{ij} + \sigma_{ij}^0 \right) \dot{u}_{i,j} \right) dV$$

$$- \lim_{\varepsilon \to 0} \int_{S_\varepsilon} \frac{1}{2} \left(\sigma_{ik} u_{i,k} \right) v_j n_j \, dS$$

Let us substitute $\sigma_{ij} = \tau_{ij} + \sigma_{ij}^0$. Then, using Hooke's law, we get

$$\dot{W} = \lim_{\varepsilon \to 0} \left\{ \int_{V - V_\varepsilon} \left(\tau_{ij} \dot{u}_{i,j} + \sigma_{ij}^0 \dot{u}_{i,j} \right) dV - \int_{S_\varepsilon} \frac{1}{2} \sigma_{ik} u_{i,k} v_j n_j \, dS \right\}$$

Using equations of motion for τ_{ij} and equations of equilibrium for σ_{ij}^0, we rewrite this equation as

$$\dot{W} = \lim_{\varepsilon \to 0}\left[\int_{V-V_\varepsilon}\left\{(\sigma_{ij}\dot{u}_i)_{,j} - \rho\dot{u}_i\ddot{u}_i + \rho f_i\dot{u}_i\right\}\,dV - \int_{S_\varepsilon}\frac{1}{2}\sigma_{ik}u_{i,k}v_j n_j\,dS\right]$$

$$(4.4.14)$$

Next we have

$$\lim_{\varepsilon \to 0}\int_{V-V_\varepsilon}\rho f_i\dot{u}_i\,dV = \frac{d}{dt}\int_V \rho f_i u_i\,dV = \frac{d}{dt}E_g \qquad (4.4.15)$$

where E_g is the gravitational energy of the volume V,

$$\lim_{\varepsilon \to 0}\int_{V-V_\varepsilon}\rho\dot{u}_i\ddot{u}_i\,dV = \frac{1}{2}\frac{d}{dt}\int_V \rho\dot{u}_i\dot{u}_i\,dV + \lim_{\varepsilon \to 0}\int_{S_\varepsilon}\frac{1}{2}\rho\dot{u}_i\dot{u}_i v_j n_j\,dS$$

$$= \frac{d}{dt}K + \lim_{\varepsilon \to 0}\int_{S_\varepsilon}\frac{1}{2}\rho\dot{u}_i\dot{u}_i v_j n_j\,dS \qquad (4.4.16)$$

where K is the kinetic energy and

$$\lim_{\varepsilon \to 0}\int_{V-V_\varepsilon}(\sigma_{ij}\dot{u}_i)_{,j}\,dV = \int_{S_0}\sigma_{ij}\dot{u}_i n_j\,dS$$

$$- \int_{\Sigma(t)}\sigma_{ij}\dot{a}_i n_j\,dS - \lim_{\varepsilon \to 0}\int_{S_\varepsilon}\sigma_{ij}\dot{u}_i n_j\,dS$$

$$(4.4.17)$$

Now instead of (4.4.14) we get

$$\dot{W} = -\dot{K} - \dot{E}_g + \int_{S_0}\sigma_{ij}\dot{u}_i n_j\,dS - \int_{\Sigma(t)}\sigma_{ij}\dot{a}_i n_j\,dS - 2\frac{d}{dt}\int_{\Sigma(t)}\gamma_{\text{eff}}\,dS$$

where equation (2.3.2) has been used for γ_{eff}. Now integrating with respect to time from 0 to t_m, we get

$$\Delta(W + E_g) + \Delta E_{\gamma_{\text{eff}}} = \int_0^{t_m}dt\int_{S_0}\sigma_{ij}\dot{u}_i n_j\,dS - \int_0^{t_m}dt\int_{\Sigma(t)}\sigma_{ij}\dot{a}_i n_j\,dS$$

$$(4.4.18)$$

where $\Delta E_{\gamma_{\text{eff}}}$ is the fracture work.

Considering equation (4.4.11) and also the fact that σ_{ij} is the frictional stress on $\Sigma(t)$, we get

$$\Delta E_p + \Delta E_{\gamma_{\text{eff}}} + \Delta E_{S_0} + A_f = 0 \qquad (4.4.19)$$

This equation describes the energy conservation law, but in contrast to (4.4.6a), it includes not only the frictional losses but also the energy dissipation due to fracture. To obtain a relation containing E_q, we transform the first term on the right side of (4.4.18):

$$\int_0^{t_m} dt \int_{S_0} \sigma_{ij} \dot{u}_i n_j \, dS = \int_0^{t_m} dt \int_{S_0} \left(\sigma_{ij} - \sigma_{ij}^0 \right) \dot{u}_i n_j \, dS + \int_{S_0} \sigma_{ij}^0 u_i^1 n_j \, dS$$

Then equation (4.4.18) can be written as

$$\Delta E_{\gamma_{eff}} + \Delta E_p = -E_q - \int_0^t dt \int_{\Sigma(t)} \sigma_{ij} \dot{a}_i n_j \, dS + \int_{S_0} \sigma_{ij}^0 u_i^1 n_j \, dS \tag{4.4.20}$$

Eliminating the change of potential energy E_p from this equation and relation (4.4.3), we obtain

$$E_q = \frac{1}{2} \int_{S_0} \left(\sigma_{ij}^0 - \sigma_{ij}^1 \right) u_i^1 n_j \, dS - \int_0^t dt \int_{\Sigma(t)} \sigma_{ij} \dot{a}_i n_j \, dS$$

$$+ \frac{1}{2} \int_{\Sigma_1} \left(\sigma_{ij}^1 - \sigma_{ij}^0 \right) a_i^1 n_j \, dS - \Delta E_{\gamma_{eff}} \tag{4.4.21}$$

We have the obvious relation

$$\int_{\Sigma_1} \sigma_{ij}^0 a_i^1 n_j \, dS = \int_0^{t_m} dt \int_{\Sigma(t)} \sigma_{ij}^0 \dot{a}_i n_j \, dS$$

using which we represent equation (4.4.21) in the form

$$E_q = \frac{1}{2} \int_{S_0} \left(\sigma_{ij}^0 - \sigma_{ij}^1 \right) u_i^1 n_j \, dS - \Delta E_{\gamma_{eff}}$$

$$- \int_0^{t_m} dt \int_{\Sigma(t)} \left(\sigma_{ij} - \sigma_{ij}^0 \right) \dot{a}_i n_j \, dS$$

$$+ \frac{1}{2} \int_{\Sigma_1} \left(\sigma_{ij}^1 - \sigma_{ij}^0 \right) a_i^1 n_j \, dS \tag{4.4.22}$$

It is seen that the seismic energy E_q does not depend on σ_{ij} or σ_{ij}^0 separately but only on the stress perturbation $\tau_{ij} = \sigma_{ij} - \sigma_{ij}^0$, as we expected above. The expression (4.4.22) for the seismic energy depends on the choice of surface S_0 and, hence, is not a characteristic parameter of the earthquake source. Dependence on S_0 is only in the first term, which contains only the static values of the final stress perturbation and displacement on S_0. The static displacements decrease with distance R from the source faster than d/R, where d is, as usual, the source dimension. Furthermore, τ_{ij} is expressed by the first derivative of u_i^1 and consequenetly decreases faster than d/R^2. Therefore, choosing for S_0 a sphere of sufficiently large radius R, we find that the first term in (4.4.22) decreases faster than d/R and may be made arbitrarily small. Neglecting this term is equivalent to the condition that seismic waves at S_0, which are used in the evulation of E_q by (4.4.1), be expressed with sufficient accuracy by the first ray approximation. If S_0 is chosen in this way $(d/R \ll 1)$, then (4.4.22) takes the form

$$E_q = -2\gamma_{eff}S + \frac{1}{2}\int_{\Sigma_1}\left(\sigma_{ij}^1 - \sigma_{ij}^0\right)a_i^1 n_j \, dS - \int_0^{t_m}dt\int_{\Sigma(t)}\left(\sigma_{ij} - \sigma_{ij}^0\right)\dot{a}_i n_j \, dS$$

$$(4.4.23)$$

where we put $\Delta E_{\gamma_{eff}} = 2\gamma_{eff}S$, S being the area of Σ_1. It is convenient to represent this expression in a somewhat different form, integrating the last term by parts with respect to time:

$$E_q = -2\gamma_{eff}S + \frac{1}{2}\int_{\Sigma_1}\left(\sigma_{ij}^0 - \sigma_{ij}^1\right)a_i^1 n_j \, dS + \int_0^{t_m}dt\int_{\Sigma(t)}\dot{\sigma}_{ij}a_i n_j \, dS$$

$$(4.4.24)$$

This is the final exact formula for the seismic energy E_q in terms of the source parameters. Following Kostrov (1974), Rudnicki and Freund (1981) have developed essentially similar relations for the energy radiated from the earthquake source.

If one assumes that γ_{eff} does not strongly depend on scale, then the term $2\gamma_{eff}S$ can be omitted for earthquakes. In fact, from the dimensional analysis, it follows that the last two terms in (4.4.24) vary with the fracture area as $S^{3/2}$, whereas the fracture work $2\gamma_{eff}S$ is proportional to S. Consequently, the relative contribution of the first term to seismic energy decreases inversely as the linear dimension of the fracture. This implies at the same time, however, that this term would significantly influence the amplitudes of acoustic pulses in laboratory studies of rock fracture, and for sufficiently small fractures, it might almost totally cancel the second term and suppress the acoustic emission, so that "silent fracture" (creep) would occur.

The second term in (4.4.24) is determined only by static quantities, that is, by the final slip and stress drop, and does not depend on the processes of fracture and sliding along the fault. At the same time, the last term contains the traction rate on the fault and strongly depends on how fracture propagates and the slip occurs during the earthquake. For example, let the interaction of the fault faces be given by the dry friction law. Then σ_{ij} would essentially be independent of time, and the last term would be absent. In that case, we would get an estimate similar to (4.3.6)

$$E_q = \frac{\Delta\sigma a S}{2} = \frac{\Delta\sigma M_0}{2\mu} \qquad (4.4.25)$$

and from (4.3.7),

$$\eta\sigma = \Delta\sigma/2 \qquad (4.4.26)$$

In this case, the apparent stress would be equal to half the stress drop.

Actually, because of the microinhomogeneity of the medium and roughness of the fault surface, the traction on the fault must vary irregularly in time, as explained in Chapters 1 and 2. In Section 3.3, the concept of incoherent radiation was associated with this irregularity of the friction and fracture propagation. It is obvious that the second term in (4.4.24) does not depend on incoherent radiation, and the third term depends basically on it since the faster the oscillations of stress, the greater will be their contribution to this term owing to the presence of the time derivative of stress. Remembering what was said in Section 2.2 about the loss of energy during slip along the fracture surfaces, we easily conclude that the third term in (4.4.24) is related to the loss of energy due to the radiation of short waves, which can be called radiational friction. In other words, we may denote

$$\Delta Q_r = -\int_0^{t_m} dt \int_{\Sigma(t)} \dot{\sigma}_{ij}(\mathbf{x}, t) a_i(\mathbf{x}, t) n_j(\mathbf{x})\, dS \qquad (4.4.27)$$

The ratio of radiation loss ΔQ_r to seismic moment can be taken as an estimate of the average radiational friction:

$$\Delta\sigma_r = \mu\frac{\Delta Q_r}{M_0} \qquad (4.4.28)$$

Thus, instead of (4.4.25) we get the standard formula

$$E_q = \frac{1}{\mu}\left(\frac{\Delta\sigma}{2} - \sigma_r\right) M_0 \qquad (4.4.29)$$

whence

$$\eta\sigma = \Delta\sigma/2 - \sigma_r \qquad (4.4.30)$$

That is, the apparent stress is equal to the difference between half the stress drop and the average radiational friction. It is obvious that ΔQ_r

takes into account only that part of the radiational loss that lies in the recorded frequency range and is not absorbed during wave propagation.

The quantity ΔQ_r depends on the correlation between the slip a_i and the traction rate $\dot{\sigma}_{ij}$. If they are uncorrelated, ΔQ_r will vanish. It must be understood that the proper representation of this term is possible only if the microstructure of the medium is taken into account, that is, beyond the framework of the macroscopic description of the medium. On the macroscale, ΔQ_r must be considered to be an independent quantity.

4.5   Seismic moment tensor and seismic flow of rocks

As established in Chapter 1, the description of the earth's material as a continuous medium depends on the choice of an elementary scale such that the same material comprises several different media at different levels of macroscopic description. So far we have considered those levels of description of a medium for which the fracture in the source of a given earthquake had macroscopic dimensions; that is, the dimensions of the particles of the medium that were considered elementary were much smaller than the dimensions of the fracture itself.

But earthquakes are not only the result of tectonic deformation. They also contribute to this deformation, merging on the whole into the process of quasi-plastic deformation. To describe this process, it is necessary to consider regions containing a large number of earthquake sources as particles of a continuous medium and time intervals much longer than the average recurrence time of these earthquakes as elementary time intervals. Such volumes and time intervals must be different for earthquakes of different magnitudes, so that at some fixed level of description large earthquakes higher than a fixed magnitude can be described discretely as fractures in a medium, and all smaller earthquakes would be smeared out and appear as deformation of this medium. Earthquakes confined to one big fault, for example, the 1906 California earthquake, represent a special case. In this case, smaller earthquakes on the whole appear as discrete sliding events on the fault, and when smeared out they represent the seismic component of slip. Historically this case was first considered by Brune (1968). The slip along the fault can be split into two parts: the seismic one, associated with the radiation of seismic waves, and the aseismic one, associated with "silent" slip, that is, creep. Obviously, this distinction depends on the recorded earthquake magnitudes since the slip due to small unrecorded earthquakes should be considered aseismic. From earthquake observations only the seismic slip can be obtained. To relate the seismic slip to seismic moments, consider

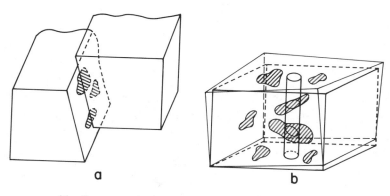

4.1   Two ways of averaging slip at the earthquake source. (a) Seismic slip; (b) seismic deformation.

as an elementary area such a part $\Delta\Sigma$ of the fault surface over which the sources (fractures) of many earthquakes during the "elementary interval of time" Δt are distributed (Figure 4.1a). Seismic slip rate along the fault at a particular place is then defined as the average slip (over these elementary areas and time intervals). Let us denote the fracture areas in these earthquakes by $\Sigma_{(\kappa)}$ $(\kappa = 1, 2, 3, \ldots, N)$ and the corresponding slips by $a^1_{i(\kappa)}(\mathbf{x})$. Then the seismic slip rate will be

$$\dot{a}_{i(q)} = \frac{1}{\Delta S \Delta t} \sum_{\kappa=1}^{N} \int_{\Sigma_{1(\kappa)}} a^1_{i(\kappa)}(\mathbf{x})\, dS = \frac{1}{\Delta S \Delta t} \sum_{\kappa=1}^{N} \bar{a}_{i(\kappa)} S_{(\kappa)}$$

$$= \frac{1}{\Delta S \Delta t} \sum_{\kappa=1}^{N} \bar{a}_{(\kappa)} b_{i(\kappa)} S_{(\kappa)} \tag{4.5.1}$$

where $\bar{a}_{(\kappa)}$ is the average slip at the κth source, $b_{i(\kappa)}$ the corresponding unit vector, and $S_{(\kappa)}$ the fracture area. Assuming the slip direction in all the earthquakes to be the same, we obtain

$$\dot{a}_{(q)} = \frac{1}{\Delta S \Delta t} \sum_{\kappa=1}^{N} \bar{a}_{(\kappa)} S_{(\kappa)} \tag{4.5.2}$$

where $\dot{a}_{(q)}$ is the seismic slip rate. Every term in this sum is the seismic moment $M_{0(\kappa)}$ of the corresponding earthquake divided by μ. Thus, the average seismic slip rate is equal to

$$\dot{a}_{(q)} = \frac{1}{\mu \Delta S \Delta t} \sum_{\kappa=1}^{N} M_{0(\kappa)} \tag{4.5.3}$$

and the seismic slip $a_{(q)}$ during time Δt would be

$$a_{(q)} = \dot{a}_{(q)} \, \Delta t = \frac{1}{\mu \, \Delta S} \sum_{\kappa=1}^{N} M_{0(\kappa)} \qquad (4.5.4)$$

Formula (4.5.3) is Brune's formula for the seismic slip rate.

Therefore, in this case the seismic moment is a measure of the contribution of an earthquake to slip along the fault, and the contribution of all the earthquakes is equal simply to the average seismic moment over the section of the fault.

In many seismically active regions, earthquakes are related to many irregularly distributed faults, so that the precision of their location is often inadequate to relate a particular earthquake to a particular fault. Such a relation may be unnecessary since the tectonic process in this case would be better described as deformation rather than as slip along a single fault. In this situation, the best description of the contribution of seismicity to tectonic deformation is presented by Riznichenko's (1965) model of seismic flow of rocks, where the fractures at the earthquake source are assumed to be randomly distributed in space.

Let us now consider the "elementary" volume (Figure 4.1b), where a large number N of earthquakes occurred in elementary time Δt with fracture surfaces $\Sigma_{(\kappa)}$ and seismic moment tensors $M_{0ij(\kappa)}$, $\kappa = 1, \ldots, N$. For the sake of simplicity, let us assume that this volume is a rectangular parallelopiped whose faces are parallel to the coordinate axes and have lengths l_1, l_2, l_3.

Take the cylindrical subvolume as shown in Figure 4.1b and consider the relative displacement of its ends. This displacement consists of the displacement due to the deformation of continuous parts of this cylinder and the sum of the displacement jumps at all the earthquake sources intersecting this elementary cylinder. The continuous deformation is of no interest to us, and the contributions from discontinuities will be given by

$$\Delta_1 u_i = \sum_{\kappa} a_{i(\kappa)} \qquad (4.5.5)$$

where the summation is carried over all the fracture areas intersecting the elementary cylinder. Let us average this displacement over all the elementary cylinders along the direction x_1. Then

$$\overline{\Delta_1 u_i} = \frac{1}{l_2 l_3} \int_0^{l_2} \int_0^{l_3} \sum_{\kappa} a_{i(\kappa)} \, dx_2 \, dx_3 \qquad (4.5.6)$$

It is convenient to convert the integrals in every term of the sum into integrals over the corresponding surface. For this we simply have to use the relation

$$dx_2\, dx_3 = n_{1(\kappa)}\, dS_{(\kappa)} \tag{4.5.7}$$

for all the fractures that intersect the elementary cylinder. Now expression (4.5.6) takes the form

$$\overline{\Delta_1 u_i} = \frac{1}{l_2 l_3} \sum_{\kappa=1}^{N} \int_{\Sigma_{(\kappa)}} a_{i(\kappa)} n_{1(\kappa)}\, dS_{(\kappa)} \tag{4.5.8}$$

Dividing by $\Delta x_1 = l_1$, we get the average distortion $\overline{\Delta_1 u_i}/\Delta x_1$ in the form

$$\frac{\overline{\Delta_1 u_i}}{\Delta x_1} = \frac{1}{\Delta V} \sum_{\kappa=1}^{N} \int_{\Sigma_{(\kappa)}} a_{i(\kappa)} n_{1(\kappa)}\, dS_{(\kappa)} \tag{4.5.9}$$

after taking into account that $l_1 l_2 l_3 = \Delta V$.

Repeating this procedure for the other two directions, we get the expression for the average distortion tensor:

$$\frac{\overline{\Delta_k u_i}}{\Delta x_k} = \frac{1}{\Delta V} \sum_{\kappa=1}^{N} \int_{\Sigma_{(\kappa)}} a_{i(\kappa)} n_{k(\kappa)}\, dS_{(\kappa)} \tag{4.5.10}$$

Since the volume ΔV is considered elementary, the divided difference can be replaced by the partial derivative (this is what the "smearing out" consists of). This gives

$$\frac{\partial u_i}{\partial x_k} = \frac{1}{\Delta V} \sum_{\kappa=1}^{N} \int_{\Sigma_{(\kappa)}} a_{i(\kappa)} n_{k(\kappa)}\, dS_{(\kappa)} \tag{4.5.11}$$

Now the increment of average deformation within the volume ΔV in time Δt would be

$$\overline{\Delta \varepsilon_{ik}} = \frac{1}{2}\left(\frac{\partial u_i}{\partial x_k} + \frac{\partial u_k}{\partial x_i} \right) \tag{4.5.12}$$

or

$$\overline{\Delta \varepsilon_{ik}} = \frac{1}{\Delta V} \sum_{\kappa=1}^{N} \int_{\Sigma_{(\kappa)}} \frac{1}{2}\left(a_{i(\kappa)} n_{k(\kappa)} + a_{k(\kappa)} n_{i(\kappa)} \right) dS_{(\kappa)} \tag{4.5.13}$$

Each term in this expression is the seismic moment tensor divided by 2μ. Thus, we get

$$\overline{\Delta\varepsilon_{ik}} = \frac{1}{2\mu\,\Delta V} \sum_{\kappa=1}^{N} M_{0ik(\kappa)} \tag{4.5.14}$$

Dividing by time Δt, we get the average strain rate due to earthquakes as

$$\dot{\varepsilon}_{(\mathrm{q})ik} = \frac{1}{2\mu} \sum_{\kappa} \frac{M_{0ik(\kappa)}}{\Delta V \Delta t} \tag{4.5.15}$$

Therefore, the average seismic strain rate tensor is equal to the sum of seismic moment tensors per unit volume.

Relation (4.5.15) is a natural generalization of Brune's formula (4.5.3) for the seismic process occurring in a volume rather than along an individual fault. Like Brune's approach, this formula can be used to calculate the long-term average seismic strain rate if an empirical relation between the seismic moment and the magnitude of earthquakes can be obtained. Such a relation should be obtained for each tectonic region. As a first approximation, one can use the relation obtained for the entire earth. Let us assume that for a volume ΔV the average orientation of sources is given by the principal axes $\mathbf{l}_{1k}$ and $\mathbf{l}_{2k}$. Then, with an accuracy sufficient for estimation purposes, the relations (4.5.15) can be rewritten, using (4.1.19) and (4.2.6), as

$$\dot{\varepsilon}_{(\mathrm{q})ik} = \frac{1}{2\mu\,\Delta V \Delta t}(\mathbf{l}_{1i}\mathbf{l}_{1k} - \mathbf{l}_{2i}\mathbf{l}_{2k}) \sum_{\kappa} M_{0(\kappa)} \tag{4.5.16}$$

Assuming the relation

$$M_0 = M_0(E) \tag{4.5.17}$$

where E is the seismic energy of the earthquake, and using the frequency–energy law,

$$dN(E) = A E^{-(\alpha+1)} \Delta E\,\Delta V \Delta t \tag{4.5.18}$$

we get the expression

$$\varepsilon_{(\mathrm{q})ik} = \frac{A}{2\mu}(\mathbf{l}_{1i}\mathbf{l}_{1k} - \mathbf{l}_{2i}\mathbf{l}_{2k}) \int_{0}^{E} M_0(E) E^{-(\alpha+1)}\,dE \tag{4.5.19}$$

for the long-term average seismic strain rate due to earthquakes in the energy interval 0 through E. Seismic observations alone are not sufficient to establish the rheological properties of the seismically flowing medium.

This is because, of all the losses of energy during seismic flow, only the seismic energy radiated is measurable. In addition to this energy, there are the fracture work and the frictional work along the fault. As a result, it seems possible to determine only that portion of the stresses in the medium that is associated with radiation (seismic viscosity) using the relation

$$\Delta E_{(q)} = \sigma_{(q)ik} \dot{\varepsilon}_{(q)ij} \Delta t \, \Delta V \tag{4.5.20}$$

where $\Delta E_{(q)}$ is the total earthquake energy in volume ΔV for the time Δt. Seismic stress tensor $\sigma_{(q)ik}$ has, by definition, the same principal axes as $\dot{\varepsilon}_{(q)ik}$, that is,

$$\sigma_{(q)ik} = \left(l_{1i} l_{1k} - l_{2i} l_{2k} \right) \sigma_{(q)} \tag{4.5.21}$$

From the frequency–energy law (4.5.18) we obtain

$$\Delta E = \frac{A}{1 - \alpha} E^{(1 - \alpha)} \Delta V \Delta t \tag{4.5.22}$$

Substituting (4.5.19), (4.5.21), and (4.5.22) into (4.5.20), we get

$$\frac{1}{(1 - \alpha)} E^{(1 - \alpha)} = \frac{1}{\mu} \sigma_{(q)} \int_0^E M_0(E) e^{-(\alpha + 1)} \, dE \tag{4.5.23}$$

whence

$$\sigma_{(q)} = \frac{\mu}{1 - \alpha} E^{(1 - \alpha)} \left[\int_0^E M_0(E) E^{-(\alpha + 1)} \, dE \right]^{-1} \tag{4.5.24}$$

If now the magnitude of the seismic strain rate

$$\dot{\varepsilon}_{(q)} = \frac{A}{2\mu} \int_0^E M_0(E) E^{-(\alpha + 1)} \, dE \tag{4.5.25}$$

is introduced, the relation of seismic resistance to the flow $\sigma_{(q)}$ and to the seismic strain rate $\dot{\varepsilon}_{(q)}$ can be studied. In particular, the coefficient of mean seismic viscosity

$$\eta_{(q)} = \sigma_{(q)} / \dot{\varepsilon}_{(q)} \tag{4.5.26}$$

can be introduced. Along with the average values defined by the frequency–energy law, it is possible to study the actual values by summing the moments of actual earthquakes. Comparison of average values with actual observations might be important in earthquake prediction. The relation (4.5.4) was recently used by Wesnousky, Scholz, and

Shimazaki (1982) to estimate the deformation of the Japan arc using seismicity and quaternary fault data to estimate $M_{0(\kappa)}$ and by Ekstrom (1987) to estimate the deformation of central Asia.

4.6 Asperity model

Consider an infinite planar fault creeping under a shear stress equal to the kinetic friction everywhere except at a finite locked patch (asperity), where stress is being accumulated. When this patch reaches a critical state, fracture is initiated and propagates across it. Suppose that the slip rate does not change sign during the process of asperity failure. Then over the whole fault the stress may be assumed unchanged except over the area of the asperity. Let us denote this area by Σ. The traction over Σ varies during the earthquake, increasing at yet unbroken parts of the asperity and dropping to the friction level behind the fracture front. Let us denote by $\tau_i(\mathbf{x}, t)$ the traction perturbation at Σ_1. Then the boundary conditions will be

$$\tau_{ij} n_j = \tau_i(\mathbf{x}, t) \qquad \text{for} \quad \mathbf{x} \in \Sigma_1$$
$$= 0 \qquad \qquad \text{otherwise} \tag{4.6.1}$$

As before, we assume that the medium can be replaced by a homogeneous, isotropic, elastic continuum. Then the resulting motion can be considered separately in the two half-spaces on the two sides of the fault. Thus, we get a boundary value problem for an elastic half-space with given tractions at its surface. The Green function for the problem is the solution to Lamb's problem (Chao, 1960; Richards, 1979). Let us denote it by $G_{ij}(\mathbf{x}, t)$. Then one obtains an equation analogous to (3.2.18):

$$\int_0^{t_1} \int_{\Sigma_1} G_{ik}(\mathbf{x}_1 - \mathbf{x}, t_1 - t) \tau_i(\mathbf{x}, t) \, dS = u_k(\mathbf{x}_1, t_1) \tag{4.6.2}$$

The expressions for G_{ik} are much more complicated than those for U_{ik} [equation (3.3.1)] and are given in Appendix 1. For the far-field asymptotics, expressions analogous to (3.3.16) and (3.3.17) can be obtained. Omitting details, we reproduce the formulas

$$u_k^{\text{P}}(\mathbf{x}_1, t' + T_{\text{P}}) = \frac{D_{ki}^{\text{P}}(\mathbf{m})}{4\pi\rho v_{\text{P}}^2 R} \int_{\Sigma_1} \tau_i\left(\mathbf{x}, t' + \frac{m_i x_i}{v_{\text{P}}}\right) dS$$

$$u_k^{\text{S}}(\mathbf{x}_1, t' + T_{\text{S}}) = \frac{D_{ki}^{\text{S}}(\mathbf{m})}{4\pi\rho v_{\text{S}}^2 R} \int_{\Sigma_1} \tau_i\left(\mathbf{x}, t' + \frac{m_i x_i}{v_{\text{S}}}\right) dS \tag{4.6.3}$$

where the directivity factors $D_{ki}^{P}(\mathbf{m})$ and $D_{ki}^{S}(\mathbf{m})$ are given by

$$D_{ki}^{P}(\mathbf{m}) = 4m_i m_k |m_3|(v_S/v_P)$$

$$\times \left\{ 4(v_S/v_P)^3 |m_s|(1 - m_3^2)\left[1 - (v_S/v_P)^2(1 - m_3^2)\right]^{1/2} \right.$$

$$\left. + \left[1 - 2(v_S/v_P)^2(1 - m_3^2)\right]^2 \right\}^{-1}$$

$$(4.6.4)$$

$$D_{ki}^{S}(\mathbf{m}) = \frac{2m_k' m_2(1 - \delta_{k3})}{(1 - m_3^2)}$$

$$- \frac{(\delta_{k3} - m_k|m_3|)m_i|m_3|(1 - 2m_3^2)\left[1 - 2m_3^2 - 4(v_S/v_P)(1 - m_3^2)\right]}{(1 - m_3^2)\left\{(1 - 2m_3^2)^2 + 4|m_3|(1 - m_3^2)\left[(v_S/v_P)^2 - 1 + m_3^2\right]^{1/2}\right\}}$$

where $\mathbf{m}' = (-m_2, m_1, m_3)$.

The directivity functions (4.6.4) are plotted in Figure 4.2, where SH denotes the S-wave component of displacement parallel to the fault plane and SV denotes the S-wave component of displacement, which lies in the plane normal to the fault plane. For comparison, the corresponding directivity functions for the dislocation problem are also shown by dashed lines. It can be seen that the sign distributions for the first arrivals of the waves coincide for both models. Thus, the models cannot be distinguished on the basis of the fault-plane solution. Note that the expressions for the SV part of u_k^S are applicable only when $m_3 < (1 - v_S^2/v_P^2)^{1/2}$ since beyond this critical angle head waves are produced along the fault surface. Then, the radiation of SV waves cannot be represented as a product of a radiation pattern with a source time function (Das and Kostrov, 1983). Hence the SV radiation pattern is plotted in Figure 4.2 only for $m_3 < (1 - v_S^2/v_P^2)^{1/2}$.

The integrals in (4.6.3) have the same form as those in (3.3.16) and (3.3.17), with the moment tensor density rate $\dot{m}_{lm}$ replaced by the dynamic stress drop τ_i. Accordingly, the whole theory of inverse problem solvability developed in Sections 3.6 and 3.7 is applicable to this model.

In contrast, the long-wave approximation considered in this chapter becomes different for the asperity model. The difference is that the stress drop τ_i tends toward a finite value $\tau_i'(\mathbf{x})$, say, after an earthquake, whereas the slip rate $\dot{a}_i$ tends toward zero. Thus, the spectra of seismic

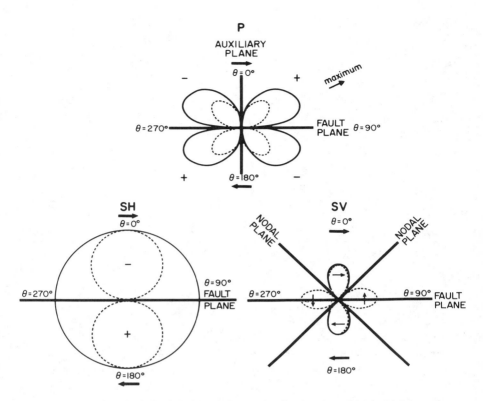

4.2   Far-field directivity functions for the asperity model (solid lines). The dashed lines are the double-couple directivity functions. The P and SV patterns for the asperity model correspond to a Poisson solid. (*After Das and Kostrov*, 1983. © *American Geophysical Union.*)

waves at low frequencies will be proportional to P_{0i}/ω, where $P_{0i} = \int_{\Sigma_1} \tau_i'(\mathbf{x}) \, dS$ is the total "force drop" at the asperity. So in this case the spectra are infinite as $\omega \to 0$. In terms of pulse shapes, this means that, whereas the dislocation model produces a one-sided displacement pulse approximated by a δ function, the asperity model produces a displacement pulse that can be approximated by the Heaviside step function. In other words, the particle velocity pulse for the asperity model is similar to the displacement pulse of the dislocation model. Instead of the seismic moment M_{0ik}, the asperity source must be characterized by the "seismic force" P_{0i}, which is proportional to the area under the particle velocity pulse due to the asperity failure. Consequently, the two models can be distinguished on the basis of their low-frequency spectra, as we shall see in Chapter 6.

These conclusions hold only for the part of a seismogram where the disturbances from the fault edge are absent. When these disturbances arrive, the dislocation (crack) model must be used. Thus, the breaking of an asperity on a finite fault produces waves with elevated low-frequency content or, in other words, longer source time function pulse, compared with the fracturing of a fault of the same size as the asperity size. This will be discussed in more detail in Part III.

Part III

SPECIFIC EARTHQUAKE SOURCE MODELS

As mentioned in the Introduction, the study of forward problems involving the earthquake source is of considerable importance, for it was and essentially remains the only way to obtain insight into the mechanics of the earthquake faulting process. The solutions to forward problems also provide constraints on the solution of inverse problems, which as we saw in Chapter 3 are generally unstable. The idealized source models considered in Chapter 5 are presented to illustrate some important observed characteristics of earthquakes; they are not representions of actual earthquake sources.

Generally applicable methods for solving dynamic crack problems analytically do not exist, and new methods have had to be invented for each particular problem. Some analytic solutions exist for self-similar problems in two (e.g., Broberg, 1960; Baker, 1962; Achenbach and Abo-Zena, 1973) and in three dimensions (Kostrov, 1964; Burridge and Willis, 1969; Willis, 1973) for planar cracks in a homogeneous medium as well as for cracks that propagate along the interface between two different media (Willis, 1971, 1972). Dynamic crack problems in which the crack tip velocity is not assigned *a priori* but is determined from some fracture criterion ("spontaneous" problems) have been considered analytically for the semi-infinite inplane and antiplane shear cracks for simple stress drop distributions on the crack (Kostrov, 1966, 1974; Eshelby, 1969; Slepyan, 1976). Freund (1972a, b, c, d, 1973, 1974) studied the dynamic propagation of tensile cracks for several cases including propagation at a constant fracture speed, as well as nonuniform propagation under various general loading conditions, and developed expressions for the energy release rate for these different cases. The treatment of the tensile problem was often useful in later solutions of shear problems since many of the techniques involved are similar. The study of self-similar problems contributed greatly to the general understanding of the way a slip on a fault is initiated and propagates during earthquakes. Since faults in the earth are finite and eventually come to a stop (thus destroying the property of self-similarity), the utility of these solutions is limited. Furthermore, the effect of the initial stress due to a preexisting crack on the dynamic solution had not been analyzed and the analytic approach could not be applied, at least for the general three-dimensional problems.

This situation motivated the development of sophisticated numerical techniques for the solution of dynamic crack problems. The advantage of numerical formulations is that the same method of solution can be applied to problems with different and arbitrary crack geometries, crack speeds, initial and loading conditions, and constitutive relations on the fault. The disadvantage is that the stress singularity (2.1.2) at the crack edge for an ideally brittle body is necessarily smeared out, and the direct application of Irwin's criterion (2.1.3), say, is not possible. One can then choose one of two conceptual approaches. Either one may apply numerical methods to the study of dynamic cracks in nonideally brittle bodies where the stress at the crack edge is required to be finite so this advantage disappears. Or one can use very fine grids in the numerical scheme to approximate adequately the stress singularity of the ideally brittle model and, for example, fit an inverse square-root curve to the stress outside the crack to find the corresponding stress intensity factor and then apply Irwin's criterion or use the J-integral approach and apply Griffith's criterion. These approaches will be discussed in detail in Section 5.6 on spontaneous fracture. With the aid of modern supercomputers, the latter approach is now feasible, at least for two-dimensional problems. Some related numerical approaches are reviewed by Atluri and Nishioka (1985).

The required fineness of the grids in the numerical method must be determined by the application of this technique to actual situations. In engineering problems in which the model parameters such as material properties and loading conditions are usually well defined, the numerical results can be compared against actual testing results to determine the adequacy of the numerical procedure. In seismological applications, however, these model parameters are essentially unknowable, at least in the sufficiently detailed form required for numerical calculations. Because of this lack of relevant information, very detailed calculations with extremely fine grids are not warranted for earthquake source models.

Our aim in Part III will be to study the general features of a dynamically propagating shear crack in order to understand the overall characteristics of the earthquake faulting process using relatively coarse grids in the numerical scheme. Obviously, even in this case, the grids must be fine enough for these general features to be properly represented.

As a result, the efficiency of the numerical algorithm is important. The finite difference and finite element methods that have been used to study dynamic shear crack problems (Andrews, 1976a, b; Madariaga, 1976; Archuleta and Frazier, 1978; Mikumo and Miyatake, 1979; Day, 1982a, b; Virieux and Madariaga, 1982) are flexible, and one could incorporate variations in medium properties, complicated problem configurations, and so on into these methods, if so desired. The price of this flexibility is loss of efficiency. In this book we shall confine ourselves to the numerical boundary–integral equation (BIE) method and solutions of planar shear cracks in infinite media. For many important problems in seismology, this method is more efficient in computer time and space, and thus permits the study of problems that are prohibitive using other numerical methods, even on supercomputers. The reason for this efficiency is that all calculations involve only quantities on the crack plane. Furthermore by comparing the results of the same problem solved by the finite difference method and the BIE method for the same grid size, Andrews (1985) found that some aspects of the dynamic solution are better represented by the latter solution. Namely, he found that the speed of waves with wavelengths on the order of the grid size are too slow in the finite difference method, thereby affecting the crack propagation speed determined. Of course, even though we confine our discussion to the BIE method, the general discussions, say on the implementation of a numerical fracture criterion or friction on fault faces, or on the discretization of the initial stress on the fault plane, are applicable mutatis mutandis to other numerical methods as well. Most important, the physical interpretation of the results of the calculations are not limited by the use of any particular method.

The BIE method has been applied to tensile crack problems, and we refer the interested reader to pages 153–8 of Kanninen and Popelar (1985) for further discussion on numerical methods used in engineering applications of these problems and for a complete bibliography on this topic.

5

THE BOUNDARY–INTEGRAL EQUATION METHOD

5.1 Representation relations

In order to reduce the earthquake source problem to the solution of a boundary–integral equation, it is necessary first to represent the stress and displacement fields throughout the medium in terms of the displacement discontinuity or the traction perturbations on the fault plane. In the case of the kinematic description of the earthquake source ("dislocation" model), such representation relations provide an explicit solution, as we saw in Section 3.2. However, as we shall show later, the boundary conditions for the dynamically described source ("crack" model) are of mixed type, and such representation relations do not give the required solution explicitly; instead, they relate the boundary values of the traction perturbations and slip, neither of which are known on the entire boundary (fault plane). Together with boundary conditions, these relations comprise the integral equations. When these integral equations are solved, the displacement and stress fields inside the body can be obtained using the same representation relations.

The geometry of the problem is shown in Figure 5.1. The plane $X_3 = 0$ is taken as the crack plane and denoted by S. Let $S(t)$ be that portion of the $X_3 = 0$ plane where slip is nonvanishing at time t and let $\bar{S}(t)$ be its complement: $S = S(t) + \bar{S}(t)$. Generally $S(t)$ is unknown and is to be determined as part of the solution. The relation between the displacement $u_k(\mathbf{X}, t)$ at any point $(\mathbf{X}, t)$ in the medium and the displacement discontinuity $a_i(\mathbf{X}', t')$ across S was given by equation (3.2.9). For a homogeneous, isotropic medium and the case of a planar shear crack, this relation becomes (with slightly different notation for convenience)

$$u_k(\mathbf{X}, t) = \int_0^t dt' \int_{S(t')} K_{\alpha 3 k}(\mathbf{X} - \mathbf{X}', t - t') a_\alpha(\mathbf{X}', t') \, dS \qquad (5.1.1)$$

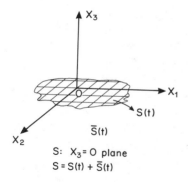

S: $X_3 = 0$ plane
S = S(t) + $\bar{S}$(t)

5.1   The geometry of the problem. $S(t)$ is the cracked portion of the $X_3 = 0$ plane.

where Latin subscripts take values 1, 2, 3 and Greek ones the values 1, 2. $K_{\alpha 3k}$ is given by

$$K_{\alpha 3k}(\mathbf{X}, t) = -\mu \left(\frac{\partial U_{k3}}{\partial X_\alpha} + \frac{\partial U_{\alpha k}}{\partial X_3} \right) \tag{5.1.2}$$

where the Einstein summation convention is assumed and U_{ik} is given by (3.3.1). Equation (5.1.1) gives the representation relation throughout the medium. To express the traction perturbation on the fault plane in terms of the slip on it, we evaluate the traction perturbation components $\tau_{\alpha 3}$ by differentiating (5.1.1) and taking the limit as $X_3 \to 0$ to obtain

$$\tau_{\alpha 3}(\mathbf{X}, t) = \int_0^t dt' \int_{S(t')} T_{\alpha\beta}(\mathbf{X} - \mathbf{X}', t - t') a_\beta(\mathbf{X}', t') \, dS \tag{5.1.3}$$

where $\mathbf{X}$ and $\mathbf{X}'$ are now the two-dimensional vectors on the fault plane S. The kernel $T_{\alpha\beta}$, obtained by differentiating $K_{\alpha 3k}$, is given by

$$T_{\alpha\beta} = -\mu \left(\frac{\partial^2 U_{33}}{\partial X_\beta \partial X_\alpha} + \frac{\partial^2 U_{\beta 3}}{\partial X_3 \partial X_\alpha} + \frac{\partial^2 U_{\alpha 3}}{\partial X_\beta \partial X_3} + \frac{\partial^2 U_{\beta\alpha}}{\partial^2 X_3} \right) \tag{5.1.4}$$

Kernel $T_{\alpha\beta}$ has strong singularities, so the integral in (5.1.3) must be considered to be a principal value. Hence, the solution of (5.1.3) is unique only under additional conditions. Usually, it is sufficient to assume that the slip a_α is smooth everywhere except at the crack edge, where it must be finite. This implies the square-root behavior of the slip a_α near the crack edge (see Section 2.3).

An alternative representation relation is obtained by exploiting the symmetries in the problem. For planar shear cracks, the solution can be shown to be antisymmetric in X_3. That is, the displacement components

u_α and traction perturbation τ_{33} are odd in X_3, whereas u_3 and $\tau_{\alpha 3}$ are even in X_3, and it is sufficient to solve the problem for the upper half-space $X_3 \geq 0$. Furthermore, from the continuity of tractions across $X_3 = 0$ [equation (1.3.25)], it follows that $\tau_{33} = 0$ everywhere on $X_3 = 0$. To obtain the required representation relation, let us first reproduce the Green–Volterra formula (3.1.5) after replacing t_1 by t and t by t' for notational convenience. Then we have

$$\int_0^t dt' \int_S (\sigma_{ij} u_i' - \sigma_{ij}' u_i) n_j \, dS + \int_0^t dt' \int_V \rho (f_i u_i' - f_i' u_i) \, dV$$

$$+ \int_V \rho (u_i \dot{u}_i' - \dot{u}_i u_i') \, dV \bigg|_{t'=0}^{t'=t} = 0 \qquad (5.1.5)$$

Let us choose as u_i' the three solutions corresponding to the three concentrated unit forces f_i' directed along the X_1, X_2, and X_3 axes and given by

$$f_i' = \delta_{ik} \delta(\mathbf{X} - \mathbf{X}') \delta(t - t') \qquad \text{for} \quad X_3' > 0, \quad X_3 \geq 0 \qquad (5.1.6)$$

where δ_{ik} is the Kronecker delta and $\delta(\mathbf{X})$ and $\delta(t)$ represent the Dirac delta function, and with the initial and boundary conditions

$$u_i' = \dot{u}_i' = 0 \qquad \text{for} \quad t \leq t'; \qquad \sigma_{i3}' = 0 \qquad \text{at} \quad X_3 = +0 \qquad (5.1.7)$$

Let us denote the solution u_i' by G_{ik}. Then in (5.1.5) the last term vanishes due to initial conditions, and the terms containing f_i and σ_{i3}' vanish due to the absence of body forces f_i and by (5.1.7), respectively. Taking σ_{ij} as the stress perturbation tensor τ_{ij} and evaluating the term with the δ function, we get

$$u_k(\mathbf{X}, t) = \int_0^t dt' \int_S G_{k\alpha}(\mathbf{X} - \mathbf{X}', t - t') \tau_{\alpha 3}(\mathbf{X}', t') \, dS \qquad (5.1.8)$$

Letting $X_3 \to 0$ and accounting for the symmetry of the displacement components, we obtain the required alternative representation relation as

$$a_\alpha(\mathbf{X}, t) = 2 \int_0^t dt' \int_S G_{\alpha\beta}(\mathbf{X} - \mathbf{X}', t - t') \tau_{\beta 3}(\mathbf{X}', t') \, dS \qquad (5.1.9)$$

where $\mathbf{X}$ and $\mathbf{X}'$ are now two-dimensional vectors on S. The required components of G are the solution to Lamb's problem and can be expressed in terms of elementary functions. The expressions for $G_{\alpha\beta}$ for the two- and three-dimensional cases are given in Appendix 1. The kernel $G_{\alpha\beta}$ possesses only weak singularities and can be directly discretized for numerical computation, as we shall discuss in Section 5.2.

Thus, we have obtained two representation relations, (5.1.3) and (5.1.9), both of which relate the slip on the crack plane to the traction perturbation on this plane. These relations comprise a set of mutually inverse integral transforms. If either the slip a_α or the traction perturbation $\tau_{\alpha 3}$ were known everywhere on S, the other of these two quantities could be obtained using (5.1.3) or (5.1.9), respectively. But in a dynamically described source (crack) problem, neither one of them is known everywhere on S. The systems (5.1.3) and (5.1.9) thus simply provide relationships between a_α and $\tau_{\alpha 3}$. Some additional relations between a_α and $\tau_{\alpha\beta}$ on the crack plane are needed to provide the required boundary–integral equations. These relations can be obtained from the constitutive relations on S, as we shall discuss in Section 5.2.

Relations (5.1.3) and (5.1.9) are equivalent in that either of them can be used together with the constitutive relations to solve the crack problem. The variables $\mathbf{X}$ and $\mathbf{X}'$ in (5.1.3) are confined to $S(t)$ since a_α vanishes outside the crack. On the other hand, the integration domain in (5.1.9) covers all points influenced by disturbances that propagate with the fastest wave velocity of the problem (e.g., the compressional wave speed of the medium for general three-dimensional problems). Therefore, one of these two relations may be more efficient than the other for a given problem. For problems in which $S(t)$ is much smaller than S, for example, "interior" crack problems, it is more advantageous to use (5.1.3). For some "exterior" crack problems, the region of traction perturbations is limited, and then (5.1.9) is the more efficient relation. Of course, the domains of integration may coincide for some particular situations – for example, for the case of a self-similar crack propagating with the fastest wave speed of the medium. The domains of integration for the two representation relations are shown in Figure 5.2 for two cases of the interior crack problem. The domain of integration for the exterior problem will be discussed in detail in Section 5.6 under "A circular asperity on an infinite fault plane."

The relation (5.1.3) was essentially formulated by Budiansky and Rice (1979). Burridge (1969) used its two-dimensional form, and Burridge and Moon (1981) its three-dimensional scalar form. The two-dimensional form of (5.1.9) was first used by Kostrov (1966) for dynamic elastic problems and by Das (1976, 1980) for general three-dimensional problems.

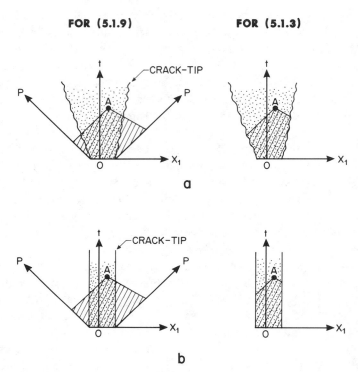

5.2   Cross sections of the domains of integration in the $(X_1 - t)$ plane for the relations (5.1.3) and (5.1.9) for (a) a propagating "interior" crack and (b) a stationary "interior" crack. The crack regions are stippled. The striped areas are the domains of integration for determining the slip at some representative point A, say, within the crack.

In the next section, we discuss the discretization of the representation relations developed here.

5.2   Discrete representations

Let us first discuss the method of discretization of relation (5.1.9). We introduce a regular network of grids centered at the points

$$X_1 = i\,\Delta X, \qquad i = -\infty, \ldots, -1, 0, 1, \ldots, \infty$$

$$X_2 = j\,\Delta X, \qquad j = -\infty, \ldots, -1, 0, 1, \ldots, \infty \qquad (5.2.1)$$

$$t = \left(k + \tfrac{1}{2}\right)\Delta t, \qquad k = 0, 1, \ldots, \infty$$

In each element of the network, we replace the tractions by their average values over the grid given by (the subscript 3 for the τ_α's being implicit from now on)

$$\tau_{\alpha ijk} = \frac{1}{\Delta t \, \Delta X^2}$$

$$\times \int_0^{\Delta t} d\tau \iint_{-\Delta X/2}^{\Delta X/2} \tau_\alpha (i \, \Delta X + \xi_1, \, j \Delta X + \xi_2, \, k \, \Delta t + \tau) \, d\xi_1 \, d\xi_2$$

$$(5.2.2)$$

We replace the slip a_α by its value at a point within the grid,

$$a_{\alpha ijk} = a_\alpha (i \, \Delta X, \, j \Delta X, \, k \, \Delta t + \delta t), \qquad 0 \le \delta t \le \Delta t \qquad (5.2.3)$$

and we replace the discrete Green function $F_{\alpha\beta}$ by

$$F_{\alpha\beta}(i, j, k) = -2 \int_0^{\Delta t} d\tau \int_{-\Delta X/2}^{\Delta X/2}$$

$$\times G_{\alpha\beta}(i \, \Delta X + \xi_1, \, j \Delta x + \xi_2, \, k \, \Delta t + \delta t - \tau) \, d\xi_1 \, d\xi_2$$

$$(5.2.4)$$

Since $G_{\alpha\beta}$ possesses integrable singularities, $F_{\alpha\beta}$ is easily evaluated using (5.2.4). The properties of $F_{\alpha\beta}$ are discussed in detail in Appendix 1. The Green function $F_{\alpha\beta}$ vanishes outside the region $v_P^2 t^2 = X_1^2 + X_2^2$. In particular, if $v_P(\Delta t - \delta t) \le \Delta X/2$, $F_{\alpha\beta}(i, j, 0)$ is nonvanishing only for $i = 0 = j$. This value is

$$F_{\alpha\beta}(0,0,0) = -2 \int_0^{\delta t} d\tau \iint_{-\infty}^{\infty} G_{\alpha\beta}(\xi_1, \xi_2, \tau) \, d\xi_1 \, d\xi_2 \qquad (5.2.5)$$

or, taking into account the homogeneity and symmetry properties of $G_{\alpha\beta}$,

$$F_{\alpha\beta}(0,0,0) = F_0 \delta_{\alpha\beta} \, \delta t \qquad (5.2.6)$$

where

$$F_0 = -2 \int_0^1 d\tau \iint_{-\infty}^{\infty} G_{11}(\xi_1, \xi_2, \tau) \, d\xi_1 \, d\xi_2 \qquad (5.2.7)$$

Here F_0 is a positive constant [the minus sign in equations (5.2.4) and (5.2.7) was included to make F_0 positive] independent of the grid size and $\delta_{\alpha\beta}$ is the Kronecker delta. The constant F_0 is the largest element in absolute value of the matrix $F_{\alpha\beta}$. Substituting the discrete forms of τ_α,

a_α, and $G_{\alpha\beta}$ into (5.1.9), we obtain

$$a_{\alpha i jk} = - \sum_{k'=0}^{k} \sum_{i'=-\infty}^{\infty} \sum_{j'=-\infty}^{\infty} F_{\alpha\beta}(i-i', j-j', k-k')\tau_{\beta i'j'k'}$$

$$+ \text{approximation error} \qquad (5.2.8)$$

In this expression, with the same order of approximation, the constant F_0 can be replaced by some other positive constant G_0, say, independent of the grid size, which is to be chosen on the basis of stability considerations.

Accordingly, (5.2.8) can be rewritten as

$$a_{\alpha i jk} + \Delta t G_0 \tau_{\alpha i jk} = - \sum_{k'=0}^{k-1} \sum_{i',\,j' \in S_{(k)}^P} F_{\alpha\beta}(i-i', j-j', k-k')\tau_{\beta i'j'k'}$$

$$\text{for} \quad i, j \in S_{(k)}^P \quad (5.2.9)$$

where $S_{(k)}^P$ is the union of all grids influenced by disturbances at time $k\,\Delta t$. The stability and approximation error of (5.2.9) are discussed in Appendix 2, where it is shown that

$$0 < \frac{F_0}{G_0} < 2 \qquad (5.2.10)$$

is a necessary condition for stability. Relation (5.2.9) is the required discrete form of the representation relation (5.1.9). For a given k, the right side of (5.2.9) depends only on the solution at previous times ($k < k'$) and (5.2.9) is an explicit scheme.

Successful discretization of equation (5.1.3), which would lead to a convenient numerical scheme, has not been achieved. Some possible approaches are discussed by Burridge (1969) and by Burridge and Moon (1981). Instead of discretizing (5.1.3), we shall use the fact that (5.1.3) is an integral transform, inverse to (5.1.9), and construct an inverse of the discrete transform (5.2.8). This inverse transform, like the direct transform (5.2.8), must be a discrete convolution transform; that is, its kernel will depend only on differences:

$$\tau_{\alpha i jk} = - \sum_{k'=0}^{k} \sum_{i'=-\infty}^{\infty} \sum_{j'=-\infty}^{\infty} S_{\alpha\beta}(i-i', j-j', k-k')a_{\beta i'j'k'}$$

$$(5.2.11)$$

Then, application of transform (5.2.8) to this kernel must give the unit kernel:

$$\sum_{k'=0}^{k} \sum_{i'=-\infty}^{\infty} \sum_{j'=-\infty}^{\infty} F_{\alpha\beta}(i-i', j-j', k-k')$$

$$\times S_{\beta\gamma}(i'-i'', j'-j'', k'-k'')$$

$$= \delta_{\alpha\gamma} \delta_{ii''} \delta_{jj''} \delta_{kk''} \qquad (5.2.12)$$

To determine $S_{\alpha\beta}$, we obtain the explicit numerical scheme from (5.2.12):

$$\Delta t G_0 S_{\alpha\gamma}(i, j, k) = -\sum_{k'=0}^{k-1} \sum_{i'=-\infty}^{\infty} \sum_{j'=-\infty}^{\infty} F_{\alpha\beta}(i-i', j-j', k-k')$$

$$\times S_{\beta\gamma}(i', j', k') + \delta_{\alpha\gamma} \delta_{i0} \delta_{j0}$$

Equation (5.2.11) can be written in a form similar to (5.2.9) as

$$\tau_{\alpha i j k} + S_0 a_{\alpha i j k} = -\sum_{k'=0}^{k-1} \sum_{i', j' \in S_{(k')}} S_{\alpha\beta}(i-i', j-j', k-k') a_{\beta i' j' k'}$$

$$\text{for} \quad i, j \in S_{(k)} \quad (5.2.13)$$

where $S_{(k)}$ is the union of all grids with nonvanishing slip and the tip element $S_0 = S_{11}(0,0,0) = 1/(\Delta t G_0)$ and is a positive constant. (The sign of F_0, G_0, and S_0 becomes important in problems where friction acts on the fault faces.) Note that using (5.2.9) is exactly equivalent to using (5.2.13) for a particular problem; that is, the solutions using the two algorithms must coincide, apart from different rounding error accumulation. The properties of $S_{\alpha\beta}$ are discussed in Appendix 2.

The approach of discretizing $T_{\alpha\beta}$ by inverting $F_{\alpha\beta}$ has two advantages. First, one can use the simple discrete representations of τ_α and a_α, that is, (5.2.2) and (5.2.3), which can be shown to be good approximations by comparing the results with analytical solutions of simple problems (an example of which is given in the next section). Direct discretization of $T_{\alpha\beta}$ may require some different representation for τ_α and a_α. Second, and more important from the practical point of view, $F_{\alpha\beta}$ is a well-behaved matrix and can be inverted without reservation. Furthermore, since $F_{\alpha\beta}$ is quite sparse, one does not actually have to invert a very large matrix [$F_{\alpha\beta}$ will consist of $(N^2 T)^2$ elements if N^2 is the number of perturbed spatial grids on the fault plane and T the maximum time level for which $S_{\alpha\beta}$ is desired] but may determine $S_{\alpha\beta}$ by an explicit time-stepping procedure.

It was mentioned in Section 5.1 that representation (5.2.13) is more economical than representation (5.2.9) for "interior" crack problems, whereas the situation is reversed for "exterior" crack problems. Let us quantify this efficiency for the numerical solution of a stationary crack – that is, a finite crack that slips without growing. In general, in seismology one is interested only in the slip distribution on the crack, because it completely determines the radiation from the crack through the asymptotic form of (5.1.3), that is, equation (3.2.11). The traction distribution outside the crack, in this context, is of interest at most in the vicinity of the crack edge. The number of grid values that are relevant at a fixed time step is proportional to the crack area in this case, whereas the number of traction values involved in algorithm (5.2.9) is proportional to the square of time. The number of arithmetic operations necessary to obtain the solution at the k th time step is proportional to k^6, and the necessary storage is proportional to k^3, when using (5.2.9). So every doubling of the number of time steps requires eight times more storage and sixty-four times longer computation time. For the stationary crack problem under discussion here, the number of slip values is proportional to k and the number of arithmetic operations when using (5.2.13) is proportional to k^2, and the storage required is proportional to k. Thus, for this problem, using (5.2.13) would be k^4 times faster and require k^2 less storage. Since the two algorithms are equally efficient for a self-similar crack propagating at the fastest wave speed of the medium, the economy in using (5.2.13) over (5.2.9) for interior problems increases with slower and slower crack speeds. Similar comparisons of efficiency can be made for exterior problems, with the above considerations applying in reverse. Algorithm (5.2.9) was first used by Hamano (1974) for two-dimensional crack problems and by Das (1980) for three-dimensional problems.

We shall now proceed to apply the algorithms developed here to some specific dynamic crack problems. In general, we shall consider only three-dimensional cases except when appropriate solutions do not exist, in which case a two-dimensional illustration will be used.

5.3   The circular self-similar shear crack

We consider the three-dimensional problem of a self-similar circular shear crack, which is initiated at a point and propagates at a known constant velocity, v say. The assumption of a constant fracture speed is rather unphysical, because it violates principles of fracture mechanics. A stress singularity that grows in time, as the stress singularity at the growing crack edge does, is unlikely to result in a constant fracture speed under any of the fracture criteria discussed in Chapter 2,

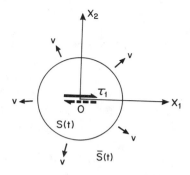

5.3   Geometry of the self-similar circular shear crack.

unless the fracture toughness distribution on the fault plane is rather pathological. Spontaneous crack problems are more physical, and we shall devote a large part of this chapter to them. It is instructive, however, to discuss the self-similar problem since it is the simplest possible case and is useful for demonstrating the numerical method, including its accuracy and stability. Historically, more than two decades ago, this problem was the first dynamic three-dimensional shear problem to be solved.

The crack region $S(t)$ is known and given by

$$S(t): X_1^2 + X_2^2 \le v_p^2 t^2$$

The geometry of the problem is shown in Figure 5.3. We shall solve the problem when the stress drop on $S(t)$ is prescribed to be a constant, $\Delta\sigma$, say.[1] Without loss of generality for the circular crack problem, we may assume that the stress drop is directed in the X_1 direction. Then we have the mixed boundary value problem

$$\tau_1 = \Delta\sigma, \tau_2 = 0 \quad \text{on} \quad S(t); \qquad a_\alpha = 0 \quad \text{on} \quad \bar{S}(t)$$

After discretization, we obtain

$$\tau_{1ijk} = \Delta\sigma, \tau_{2ijk} = 0$$

$$\text{for} \quad S_{(k)}: (i\,\Delta X)^2 + (j\,\Delta X)^2 \le v^2 \left(\frac{k+1}{2}\right)^2 (\Delta t)^2 \quad (5.3.1)$$

$$a_{\alpha ijk} = 0 \quad \text{for} \quad \bar{S}_{(k)}: (i\,\Delta X)^2 + (j\,\Delta X)^2 > v^2 \left(\frac{k+1}{2}\right)^2 (\Delta t)^2$$

---

[1] The term "stress drop" has traditionally been used in seismology to mean traction drop, and we shall continue to use the term in this sense.

Taken together with (5.2.9) or (5.2.13) this gives the complete formulation of the problem.

We first use (5.2.9) to solve the problem. We denote its right side by, say, $L_{\alpha i j(k-1)}$, that is,

$$L_{\alpha i j(k-1)} = - \sum_{k'=0}^{k-1} \sum_{i',\, j' \in S_{(k')}^{P}} F_{\alpha\beta}(i - i',\, j - j',\, k - k')\tau_{\beta i' j' k'}$$

$$\text{for} \quad i,\, j \in S_{(k)}^{P}$$

where $S_{(k)}^{P}$ was defined in the last section. The term $L_{\alpha i j(k-1)}$ depends on the values of τ_α at all previous time steps and is thus known at any time step if the traction history up to the previous time step is known. The summation in $L_{\alpha i j(k-1)}$ extends over the entire cone of dependence of the grid point $(i \, \Delta X,\, j\Delta X,\, (k + \frac{1}{2}) \, \Delta t)$ except the grid point itself. Then we can rewrite (5.2.9) as

$$a_{\alpha i j k} + \Delta t G_0 \tau_{\alpha i j k} = L_{\alpha i j(k-1)}$$

It follows very simply from this that, under the mixed boundary conditions (5.3.1), the solution is given by

$$\begin{aligned} a_{\alpha i j k} &= -\delta_{\alpha 1} \Delta t G_0 \, \Delta\sigma + L_{\alpha i j(k-1)} \\ \tau_{\alpha i j k} &= \delta_{\alpha 1} \, \Delta\sigma \quad \text{on} \quad S_{(k)} \end{aligned} \qquad (5.3.2)$$

and by

$$\tau_{\alpha i j k} = \frac{L_{\alpha i j(k-1)}}{\Delta t G_0}; \qquad a_{\alpha i j k} = 0 \quad \text{on} \quad \bar{S}_{(k)} \qquad (5.3.3)$$

where $\delta_{\alpha 1}$ is the Kronecker delta. Since the initial $L_{\alpha i j(k-1)}$ is zero by definition, this is an explicit scheme to determine slip and traction perturbation everywhere on the crack plane.

Let us now solve the same problem using the discrete representation (5.2.13). We denote its right side by, say, $M_{\alpha i j(k-1)}$, that is,

$$M_{\alpha i j(k-1)} = - \sum_{k'=0}^{k-1} \sum_{i',\, j' \in S_{(k')}} S_{\alpha\beta}(i - i',\, j - j',\, k - k')a_{\beta i' j' k'}$$

$$\text{for} \quad i,\, j \in S_{(k)}$$

where $S_{(k)}$ was defined earlier. The term $M_{\alpha i j(k-1)}$ depends on the values of a_α at all previous time steps and is thus known at any time step if the slip history up to the previous time step is known. The summation in $M_{\alpha i j(k-1)}$ extends over the intersection of the cone of dependence of the point $(i \, \Delta X,\, k \, \Delta X,\, (k + \frac{1}{2}) \, \Delta t)$ with the crack area $S_{(k)}$ but excluding this point itself, that is, over a smaller region than when (5.2.9) was used. We can now rewrite (5.2.13) as

$$\tau_{\alpha i jk} + S_0 a_{\alpha i jk} = M_{\alpha i j(k-1)}$$

The solution to this under the mixed boundary conditions (5.3.1) is very simply

$$\tau_{\alpha i jk} = \delta_{\alpha 1} \Delta\sigma; \qquad a_{\alpha i jk} = -\frac{\delta_{\alpha 1} \Delta\sigma + M_{\alpha i j(k-1)}}{S_0} \qquad \text{on} \quad S_{(k)}$$

$$(5.3.4)$$

This is an explicit scheme for finding $a_{\alpha i jk}$ on $S_{(k)}$ since $M_{\alpha i j(k-1)}$ is known to be zero initially and gives the required slip on the crack. If the solution for the stress $\tau_{\alpha i jk}$ on $\bar{S}(k)$ is desired, it can be easily obtained as

$$a_{\alpha i jk} = 0; \qquad \tau_{\alpha i jk} = M_{\alpha i j(k-1)} \tag{5.3.5}$$

It is important to note that the solutions for the traction perturbation and slip are homogeneous functions of zeroth order of the coordinates and time for self-similar problems (definition of self-similarity).

We shall now compare the numerical solution using the two forms of the representation relations with the corresponding analytic self-similar solution. Let us consider the case in which the fracture velocity $v = v_P/2$. The results are shown in Figure 5.4 for the case of a Poisson solid. (In the remainder of the book all numerical results will be illustrated for the Poisson solid only.) The required analytic solution for the slip a_α is obtained by integrating equation (3.4.18) with respect to time and substituting $v = v_P/2$. The numerical solutions are determined from (5.3.2), (5.3.3) and from (5.3.4), (5.3.5) for the two forms of the representation relations. The two numerical methods yield solutions that are identical except at the last decimal place, as expected. Since the initial crack in a numerical method cannot be infinitesimal, one cannot in fact numerically study a self-similar crack. In this example, an initial crack of radius ΔX is assumed to appear instantaneously and start extending at a speed of $v_P/2$. In other words, the analytic and numerical solutions are really solutions of different problems. However, at larger and larger times, the effect of this initial difference is expected to become less and less significant. The normalized half-slip U is plotted in units of $(20 \Delta\sigma \Delta X)/3\mu$ and the normalized time T is plotted in units of $v_P/\Delta X$. The analytic solution for the half-slip is shown by the continuous line in Figure 5.4; its numerically calculated values are shown by crosses at

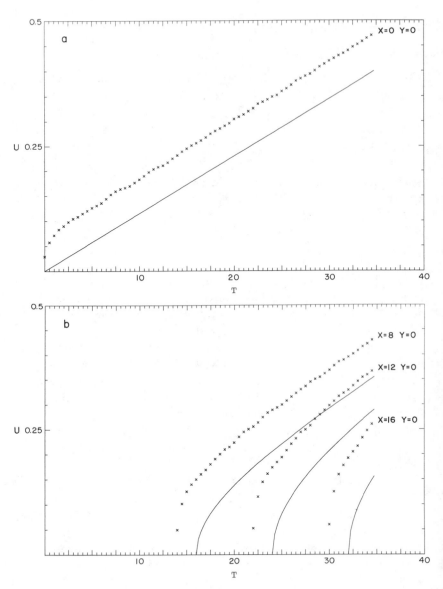

5.4   Normalized half-slip U versus normalized time T for a circular crack
propagating at a speed $v_P/2$. The analytic solution is indicated by solid lines,
and the numerical solution for the instantaneously appearing circular crack of
radius ΔX is indicated by crosses. $X = X_1/\Delta X$; $Y = X_2/\Delta X$. The solution is
plotted at points (a) and (b) along the X_1 axis, (c) along the X_2 axis, (d) along a
line at 45° to the axes [the analytic solution is omitted to increase clarity in (d)].
The spatial to temporal grid size ratio is given by $v_P \, \Delta t/\Delta X = .5$.

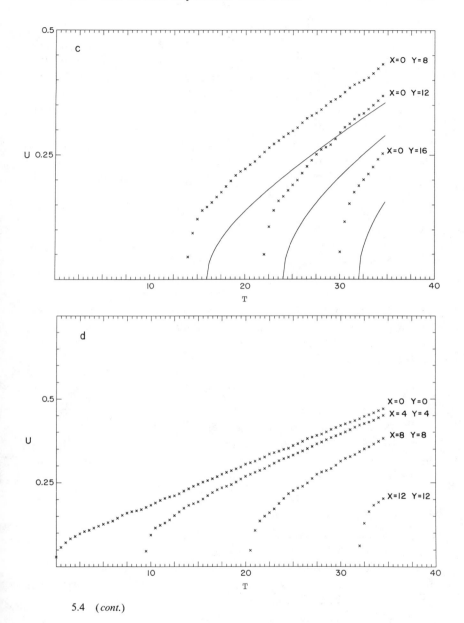

5.4   (*cont.*)

points along the X_1 and X_2 axes and along a line at 45° to the two axes. The numerical solution always lies above, that is, is larger than the analytic solution owing to the above-mentioned initial difference. What is most important, however, is that the *rate* of increase of slip with time is the same for the analytic and numerical solutions except for the first few time steps of the solution. To ensure stability, G_0 was taken as $2F_0$, which satisfies the necessary criterion for stability given by (5.2.10). We note that the square-root form of the slip a_1 is well approximated numerically. The slip component a_2 on the crack was always found to be less than 3 percent of a_1; in other words, it is practically negligible (it is identically zero in the analytic case). In fact, the maximum values of a_2 are concentrated in grids near the crack edge, which in a numerical scheme is necessarily smeared out, and in the interior of the crack its values are even smaller. The values of a_2 can be considered a measure of the numerical noise in the solution. Thus, the slip is in the direction of the stress drop on the fault plane for the circular self-similar crack. Figure 5.4 also shows that the numerically determined slip is azimuthally symmetric (as the analytic one is) without the *a priori* imposition of such a condition. The three-dimensional self-similar circular shear crack problem was numerically studied by Madariaga (1976), Archuleta (1976), and Das (1980), among others.

So we have used the numerical boundary–integral method in its two forms to study a simple self-similar problem and showed that the results compare well with its analytic solution for one particular fracture speed. In a similar way, any crack problem with given stress drop on the fault plane and crack speed (neither of these need be constant and the speed need not be the same in all directions) can be solved following the above development. It includes as a special case the stationary crack problem studied by Madariaga (1976) and by Das (1980).

5.4   The finite circular shear crack

Next we consider the problem of a circular shear crack that initiates at a point, propagates at a preassigned constant velocity v, say, and stops when it reaches some finite radius r, say. Let the stress drop on the crack be assigned a constant, $\Delta\sigma$, say, and directed in the X_1 direction. The crack region $S(t)$ is defined by

$$S(t): X_1^2 + X_2^2 \leq v^2 t^2 \qquad \text{for} \quad vt \leq r;$$

$$X_1^2 + X_2^2 = r \qquad \text{for} \quad vt > r$$

$$(5.4.1)$$

Then we have the mixed boundary value problem

$$\tau_1 = \Delta\sigma, \tau_2 = 0 \quad \text{on} \quad S(t); \qquad a_\alpha = 0 \quad \text{on} \quad \bar{S}(t)$$

$$(5.4.2)$$

This problem can be solved numerically following the procedure outlined in detail in the last section, the discrete crack area $S_{(k)}$ now being defined as

$$S_{(k)} : \begin{cases} (i\,\Delta X)^2 + (j\Delta X)^2 \le v^2\big(k + \tfrac{1}{2}\big)^2(\Delta t)^2 \\[2mm] \hspace{4cm} \text{for} \quad v\big(k + \tfrac{1}{2}\big)\,\Delta t \le r \\[2mm] (i\,\Delta X)^2 + (j\Delta X)^2 = r^2 \quad \text{for} \quad v\big(k + \tfrac{1}{2}\big)\,\Delta t > r \end{cases}$$

Even this relatively simple problem of a finite dynamic crack cannot be solved analytically, though a kinematic description using the results of the self-similar problem was considered by Sato and Hirasawa (1973).

Let us apply the numerical algorithm (5.2.13) to this problem. We consider an instantaneously appearing crack of diameter $3\,\Delta X$, which grows to a final diameter of $41\,\Delta X$ at a speed $v = v_\mathrm{P}/2$. We shall allow backslip to occur on the crack in this example. The half-slip $a_1/2$ is plotted against time in Figure 5.5 and is normalized by $(r\Delta\sigma)/3\mu$ for this problem. The slip on the crack is found to be essentially in the direction of the stress drop even after the crack has stopped, and it coincides with the solution of the corresponding self-similar problem until the first diffracted waves from the crack edge arrive and decrease the slip rate. The rise time and slip at the center are larger than at the crack edge. The dynamic slip is found to overshoot the static value, but when backslip is allowed on the crack, as it is here, it decreases from its maximum value and approaches the static value.

From some simple geometric considerations, it is possible to obtain a rough estimate of the dynamic overshoot expected at a point within the crack. A point on the crack continues slipping until some (diffracted) wave from the crack edge returns to it. Let this wave have velocity v_H. Then the displacement expected at a point on the crack is given by the solution of the dynamic self-similar circular crack propagating at the speed v. The overshoot OV on the crack varies with position on the crack, being largest at the center and smallest at the edge. At the center, the overshoot is given by

$$OV \approx \frac{A(v)}{A(0)}\left(1 + \frac{v}{v_\mathrm{H}}\right) - 1$$

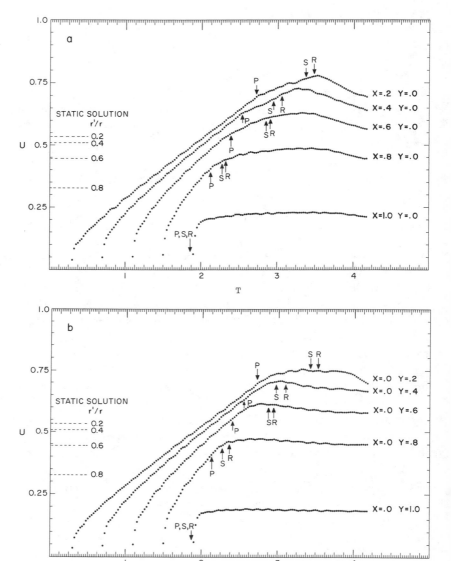

5.5 Normalized half-slip U versus normalized time $T = v_P^t/r$ for a finite circular crack for (a) points along the X_1 axis and (b) points along the X_2 axis. $X = X_1/r$, $Y = X_2/r$, r being the final crack radius, and $v_P \Delta t/\Delta X = .5$. The theoretical arrival times of the diffracted P, S, and Rayleigh waves from the crack edge are indicated on each curve by arrows. The static solutions for the points $r'/r = .2$, .4, .6, and .8, where r' is the distance of the point from the center of the crack, are given by $U = .55\sqrt{1 - r'^2/r^2}$ and marked along the abscissas.

where $A(v)/A(0)$ was plotted in Figure 3.6 as a function of v. For $v = v_P/2$, this overshoot is $0.85(1 + v_P/2v_H) - 1$. A lower bound of this overshoot would be obtained for $v_H = v_P$, that is, if the wave that stops the slip is the P wave. This lower bound at the center of the crack is ~ 28 percent above the corresponding static slip from the above formula. Figure 5.5 shows that the maximum slip is reached at or soon after the time when the Rayleigh wave arrives from the crack edge. Then, taking $v_H = v_R$, the Rayleigh wave velocity, one obtains a very rough estimate of the upper bound of the overshoot at the crack center as ~ 65 percent. This is a very rough estimate since it assumes that the P and/or S waves from the crack edge did not modify the slip determined by the dynamic self-similar solution (an assumption that is seen from Figure 5.5 to be not quite valid!). At the crack edge the overshoot is

$$OV \approx \frac{A(v)}{A(0)} \sqrt{1 + v/v_H} - 1$$

so that for $v_H = v_P$, the edge overshoot is ~ 4 percent, and for $v_H = v_R$, it is ~ 18 percent of its static value. The static values of U are shown in Figure 5.5, and the central overshoot in the numerical case was found to be ~ 36 percent. Estimates of this overshoot, obtained by Madariaga (1976), Archuleta (1976), Das (1980), and others using different numerical methods and/or different grid sizes and without allowing backslip were found to lie between 20 and 27 percent. The dynamic overshoot of slip in the interior of the crack may, of course, be interpreted as the overshoot of the static stress drop there. The above formulas also show that the dynamic overshoot increases with increasing crack speed v. Obviously for a given problem (solved by the same method and using the same discretization of the problem), the overshoot must be greater when backslip is disallowed on the fault. Hence, in obtaining an estimate of the static solution, allowing backslip would give a better estimate than disallowing it.

5.5 A crack with dry friction

We shall consider the same crack as in the previous section but with dry friction acting on its surfaces. If the slip rate at a point on the crack surface does not vanish, the total shear traction is determined by equation (1.3.28). Or, denoting the product of the coefficient of friction and the normal stress by σ^{kin}, it is given by

$$\sigma_\alpha = -\sigma^{kin} a_\alpha/\dot{a} \qquad \text{when} \quad \dot{a} \neq 0 \tag{5.5.1}$$

That is, the traction is collinear with the slip rate, and its magnitude is equal to the kinetic frictional traction. In what follows, we shall denote the magnitude of a two-dimensional vector by the same letter without subscript, namely, $\dot{a} = \sqrt{\dot{a}_\alpha \dot{a}_\alpha}$, $\sigma = \sqrt{\sigma_\alpha \sigma_\alpha}$, and so on. As long as the crack continues to propagate, the solution to this problem coincides with the solution to the corresponding self-similar problem discussed in an earlier section. Indeed, the slip and slip rate directions coincide with the direction of the initially applied stress σ_α^0 (assumed to be constant) and the magnitude of the stress drop $\Delta\sigma = |\sigma^{\mathrm{kin}} - \sigma^0|$, it also being in the same direction as σ_α^0. If we assume that the initial stress is applied in the X_1 direction, we have the boundary conditions (5.4.2) with the above stress drop $\Delta\sigma$. When the crack stops propagating, the boundary conditions become more complicated. As long as a point on the crack continues to slip, relation (5.5.1) holds, but now the direction of the slip velocity does not necessarily coincide with that of the initial stress. If, however, the slip rate at a point on the crack vanishes, the point becomes locked and can resume sliding only if the total shear stress reaches a static friction value σ^{stat}, which is assumed to be greater than or equal to σ^{kin}, so that

$$\dot{a} = 0 \quad \text{when} \quad \sigma \leq \sigma^{\mathrm{stat}}, \quad \text{where} \quad \sigma^{\mathrm{stat}} \geq \sigma^{\mathrm{kin}} \qquad (5.5.2)$$

Consequently, every grid on the crack surface may be in one of two different states, namely, in the locked state when the condition (5.5.2) holds or in the sliding state corresponding to condition (5.5.1). Condition (5.5.1) involves only the direction of the slip rate vector, which in a numerical representation can be replaced by the slip increment $\Delta a_{\alpha i jk} \equiv (a_{\alpha i jk} - a_{\alpha i j(k-1)})$. Then condition (5.5.1) can be rewritten as

$$\sigma_{ijk} = \sigma^{\mathrm{kin}}; \quad \left(\sigma_{ijk} \text{ and } \Delta a_{\alpha i jk} \text{ have the same direction}\right)$$
$$(5.5.3)$$

where $\sigma_{\alpha i jk} = \tau_{\alpha i jk} + \tau_\alpha^0$ and σ_{ijk} is its magnitude. It is convenient to rewrite the representation expressions (5.2.9) and (5.2.13) in the generalized form

$$A\tau_{\alpha i jk} + Ba_{\alpha i jk} = R_{\alpha i j(k-1)} \qquad (5.5.4)$$

where $A = \Delta t G_0$, $B = 1$ for the relation (5.2.9) and $A = 1$, $B = S_0$ for the relation (5.2.13). Here $R_{\alpha i j(k-1)}$ denotes the corresponding right-hand sides of these two relations, which depend only on quantities at previous time steps and, for any k, have already been evaluated. Both A and B are positive constants. Let us now express the last equation in terms of the total stress σ_α and the slip increment Δa_α, obtaining

$$A\sigma_{\alpha i jk} + B\Delta a_{\alpha i jk} = N_{\alpha i j(k-1)} \tag{5.5.5}$$

where

$$N_{\alpha i j(k-1)} = R_{\alpha i j(k-1)} + A\sigma_\alpha^0 - Ba_{\alpha i j(k-1)} \tag{5.5.6}$$

which is again known from values at previous time steps. From the second condition of (5.5.3) and equation (5.5.5), it follows that $N_{\alpha i j(k-1)}$ has the same direction as both $\sigma_{\alpha i jk}$ and $\Delta a_{\alpha i jk}$. Consequently, equation (5.5.5) holds for their magnitudes as well:

$$A\sigma_{ijk} + B\Delta a_{ijk} = N_{ij(k-1)} \tag{5.5.7}$$

If a grid was sliding at the previous time step and continues sliding, then according to (5.5.3), $\sigma_{ijk} = \sigma^{kin}$, which is possible only if $N_{ij(k-1)} > A\sigma^{kin}$. In this case,

$$\Delta a_{ijk} = \frac{1}{B}\left(N_{ij(k-1)} - A\sigma^{kin}\right)$$

Since the direction of $\Delta a_{\alpha i jk}$ coincides with that of $N_{\alpha i j(k-1)}$, we find that

$$\Delta a_{\alpha i jk} = \frac{1}{B}\left(1 - \frac{A\sigma^{kin}}{N_{ij(k-1)}}\right)N_{\alpha i j(k-1)}$$

Now from (5.5.5), we obtain

$$\sigma_{\alpha i jk} = \frac{\sigma^{kin}}{N_{ij(k-1)}}N_{\alpha i j(k-1)}$$

Finally, for $a_{\alpha i jk}$ and $\tau_{\alpha i jk}$, we find

$$a_{\alpha i jk} = \frac{1}{B}\left(1 - \frac{A\sigma^{kin}}{N_{ij(k-1)}}\right)N_{\alpha i j(k-1)} + a_{\alpha i j(k-1)}$$

$$\tau_{\alpha i jk} = \frac{\sigma^{kin}N_{\alpha i j(k-1)}}{N_{ij(k-1)}} - \sigma_\alpha^0 \tag{5.5.8}$$

which gives the solution for a previously sliding grid when $N_{ij(k-1)} > A\sigma^{kin}$.

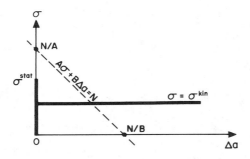

5.6   Schematic representation of the solution for a sliding grid for $N_{ij(k-1)} >$ $A\sigma^{kin}$

If $N_{ij(k-1)} \leq A\sigma^{kin}$, further sliding becomes impossible and the grid becomes locked. Then, $\Delta a_{\alpha ijk} = 0$ and we have

$$a_{\alpha ijk} = a_{\alpha ij(k-1)}$$

$$\tau_{\alpha ijk} = \frac{N_{\alpha ij(k-1)}}{A} - \sigma_{\alpha}^0 \tag{5.5.9}$$

Consider now the case of a previously locked grid. If $N_{ij(k-1)} < A\sigma^{stat}$, then the grid remains locked and solution (5.5.9) holds. In contrast, if $N_{ij(k-1)} \geq A\sigma^{stat}$, then the grid becomes unlocked and we again obtain solution (5.5.8).

The solution for $N_{ij(k-1)} \geq A\sigma^{kin}$ (i.e., for a sliding grid) is graphically represented in Figure 5.6, where the friction law is shown by a thick line and relation (5.5.7) is represented by a dashed line. The slope of this straight line is equal to $-1/(\Delta t G_0)$ and is determined by the magnitude of Δt (since G_0 is independent of the grid sizes, as we saw in Section 5.2). The finer the grid, the steeper is the slope of this line. We shall call the quantity $-1/(\Delta t G_0)$ the "local stiffness." (Andrews, 1985, called its reciprocal the "local compliance.") Note that this slope tends toward infinity as Δt tends toward zero.

One possible generalization of the dry friction model is the slip-weakening friction model (Rice, 1980). In this model, the stress does not suddenly drop from the static friction level σ^{stat} to σ^{kin} but decreases during some preliminary slip, eventually reaching σ^{kin}. In the simplest case, the frictional stress decays linearly with the amount of the pre-liminary slip. This model is schematically shown in Figure 5.7, where the quantity a_0 is called the "critical" slip-weakening displacement. If the slope of the slip-weakening friction curve is greater than the local stiffness, it is easily seen from the figure that our previous

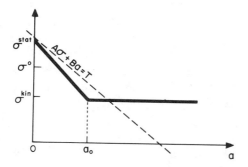

5.7 Schematic representation of the linear slip-weakening friction model.

solution, that is, equations (5.5.8) and (5.5.9), holds, because the straight line

$$A\sigma_{\alpha i jk} + Ba_{\alpha i jk} = T_{\alpha i j(k-1)}$$

where

$$T_{\alpha i j(k-1)} = R_{\alpha i j(k-1)} + A\sigma_{\alpha}^0$$

has a slope of $-1/\Delta t G_0$ and does not intersect the inclined part of the friction curve. In other words, the slip-weakening process is completed within a single grid during a single time step. Consequently, for insufficiently fine grids, the slip-weakening model of friction is equivalent to that of Coulomb dry friction.

5.6 Spontaneous cracks

In the seismological context, cracks must be considered as initiating, propagating, and arresting spontaneously, just as earthquakes do. For this, one must include in the problem formulation a finite, initial (preexisting) crack along with the initial stress distribution around it as well as some fracture criterion.

Initial stress on the fault

In the earth, the initial stress distribution on a fault before an earthquake is determined not only by the stress drop history on the fault over several earthquake cycles, but also by the preparatory processes on the fault including foreshocks, by the fluid diffusion around the fault, and so on. The actual initial stress on the fault is thus rather inhomogeneous and in practice may be unknowable. Here we shall consider only the perturbation of an initial uniform stress by the presence of a single initial crack (or asperity). Following this approach the initial stress due to multiple cracks or asperities can be determined.

For the numerical solution of the problem, it is necessary first to produce the discrete representation of the initial stress distribution on the fault. In the simplest cases when a single initial crack has a circular or elliptical form, analytic solutions to the corresponding static problems are available. The discretization of these analytic solutions is, however, not quite trivial. If the analytic initial stress is simply averaged over spatial grids, the resulting values at grids at the edge of the crack behave chaotically; that is, the average stress may be very high in one grid and very low in the adjacent grid, depending on the portion of the grid intersecting the initial crack area. To overcome this difficulty, the initial stress values at the edge grids must be artificially assigned to be equal to the average stress over an equivalent area more optimally situated relative to the initial crack edge. One such approach was taken by Das and Kostrov (1983, 1985). The resulting behavior of the solution may be influenced by a particular choice of these artificially assigned values. A more consequent approach may be to consider the instantaneous appearance of the initial crack and to apply algorithm (5.2.9) or (5.2.13), whichever is more efficient for the case being considered, to determine the traction distribution on the fault. For this the crack edge must be held stationary and the problem solved for a sufficiently long time to reach the final quiescent state. Then this traction distribution may be used as an initial condition for the spontaneously propagating crack. The advantage of this approach is that the initial stress state for problems for which analytic solutions do not exist, for example, cracks and asperities having shapes other than circular or elliptical, preexisting multiple cracks or asperities, can be easily obtained and included in the formulation of the dynamic problem. This method will be used in one of the examples considered in this section.

Fracture criterion

Fracture criteria for continuous, ideally brittle bodies are essentially nonlocal. Even though we referred to Irwin's criterion in Chapter 2 as a local criterion, closer examination of it shows that, since it is formulated in terms of a stress intensity factor, which reflects how steeply the stress increases when approaching the crack edge, it depends on the distribution of stress in a (infinitesimal) region surrounding the crack edge. Consequently, the state of a point (broken or unbroken) depends on the state of nearby points. To simulate this criterion numerically, one may approximate the stress distribution outside the crack by a function that describes the inverse square-root dependence of the stress on the distance from a curve that approximates the crack edge, the latter to be determined as part of this fitting procedure (Kanninen and Popelar, 1985). The stress intensity distribution along this curve can then be used in the fracture criterion formulation. Obviously, this fitting procedure for the stress should include only unbroken grids and exclude neighboring broken grids, where the values of stress would not be representative. Alternatively, one could approximate the square-root behavior of the slip inside the crack or the behavior of both the slip inside the crack and the stress outside it simultaneously. To apply such a procedure, it is necessary to have rather fine grids, with the spatial grid size being much less than the crack size, say. An attempt to approximate the inverse square-root behavior outside the crack using relatively coarse grids was made by Virieux and Madariaga (1982). For three-dimensional problems, such an approach is just now becoming practical with the rapid advance in supercomputer capabilities but has not yet been undertaken. The excellent fit of the numerically calculated stresses outside the crack to an inverse square-root form for two-dimensional problems using very fine grids (Andrews, 1985) indicates that this approach holds promise for three-dimensional problems.

We now proceed from a discussion of nonlocal fracture criterion to a discussion of truly local fracture criteria and their numerical formulation. It is impossible to represent the ideally brittle body within a discrete model because the transition from the unbroken to the broken state occurs within the finite grid and during a finite time step. Since the details of this transition may be masked within a single grid, one has no means to make the transition instantaneous and concentrated in space. In analogy with the friction model discussed in the last section, when the transition zone becomes smaller than the spatial grid size and the transition takes less than a time step, the linear slip-weakening fracture model (see Chapter 2) produces the classical critical stress or strength fracture criterion. Obviously, this conclusion is applicable to other non-ideally brittle models. Accordingly, the discrete scheme of fracture using the classical strength criterion can be called the "discrete ideally brittle model" because, as in the case of the continuous ideally brittle body, the details of the transition from broken to unbroken are neglected.

If one chooses to consider the discrete model as an approximation to the continuous ideally brittle model, it is necessary to force the solutions of discrete crack problems to tend toward the solution of the continuous problem when the grid size tends toward zero. Alternatively, one may wish to force the discrete solution to behave like the continuous one at a fixed grid size, for example, to prohibit the crack propagation speed from being greater than the limits inferred from the continuous ideally brittle body model. Of course, if the numerical solution is forced to tend toward that of the continuous problem, then for sufficiently fine grids the discrete crack behavior will automatically be similar to the continuous one. Obviously, in the first instance one must make the critical stress level depend on the grid size. It can be simply shown that the critical stress level must increase proportionally to the inverse square root of the grid size, the proportionality factor itself being proportional to the critical stress intensity factor (Das and Aki, 1977a). In the second instance, to simulate the crack behavior one can try to introduce a model of the transition process within the spatial and temporal grid. This approach is called the singular element method and has been applied to static crack problems in shear and tension and to two-dimensional dynamic crack problems in tension. [The literature on this latter topic is so vast that we simply refer to the recent review of Atluri and Nishioka (1985) and to the textbook of Kanninen and Popelar (1985, pp. 53–4, 153–8) for a complete bibliography and detailed discussions.] The above arguments lead to the conclusion that to simulate numerically the ideally brittle continuous body, one has to apply the nonideally brittle discrete model!

To illustrate the general features of the numerical treatment of the fracture criterion, let us consider a linear slip-weakening type of behavior as described in the previous section but with the static frictional level σ^{stat} replaced by the critical stress say, σ^Y, and with a different critical slip, say a^Y. For the purpose of the discussion here, we shall restrict ourselves to the case when σ^Y and a^Y are constant in space and time. In contrast to the previous section, we will then have grids in *three* possible states: unbroken, broken and sliding, broken and locked. For the unbroken grids we then have relations of the form (5.5.9), that is,

$$a_{\alpha i j k} = 0$$

$$\tau_{\alpha i j k} = \frac{1}{A} N_{\alpha i j (k-1)} - \sigma_\alpha^0$$

where σ^0 is the initial stress on the crack plane. If $N_{\alpha i j (k-1)}$ so calculated is greater than or equal to $A\sigma^Y$, the grid is allowed to break and is included into the crack area. The slip and traction perturbation at this grid will be found from the slip-weakening curve for fracture and equation (5.5.7). The behavior of grids within the crack must be analyzed using the slip-weakening friction curve, as discussed in the previous section. If the grid is so coarse that the local stiffness is greater than the slope of both the slip-weakening curves for fracture and friction, the solutions given in (5.5.8) and (5.5.9) are applicable. In this case the slip-weakening fracture criterion produces the classical strength criterion. This numerical criterion is truly local in the sense that the state of a grid at a given time does not depend on the state of adjacent grids at the *same* time! The two-dimensional analog of this was called the "critical stress level" fracture criterion by Das and Aki (1977a).

Crack arrest

The conditions for crack arrest were discussed in Chapter 2. In the numerical scheme, a spontaneous crack will arrest when $N_{ij(k-1)} < A\sigma^Y$. This will occur if σ^Y is too large or if $N_{ij(k-1)}$ is too small; that is, the crack will arrest when the strength of the material to be fractured is too high or the stress drop on the crack is too low. In other words, cracks arrest because of an inhomogeneous distribution of strength and stress drop on the fault (see Husseini et al., 1975), but we will not go into this problem in detail here. In the examples below, we shall simply stop the crack by placing extensive unbreakable regions ("barriers") on the fault.

For the final set of examples in this chapter, we shall consider several cases of spontaneous cracks propagating under the "critical stress level" fracture or friction criterion and one case of a spontaneous crack under the linear slip-weakening criterion.

Examples

Spontaneous shear cracks propagating on infinite planes of constant strength: We consider an instantaneously appearing shear crack of finite diameter that grows spontaneously on an infinite plane of constant critical stress σ^Y or constant static frictional limit σ^{stat} and has uniform assigned stress drop $\Delta\sigma$, say. This implies that the numerical results can be interpreted both in terms of a spontaneous fracture model and in terms of a dry friction model. We shall use σ^u to denote either of the quantities σ^Y or σ^{stat}. Laboratory experiments on rocks suggest that usually $\sigma^Y > \sigma^{stat}$. Without loss of generality for this problem, we can consider the stress drop $\Delta\sigma$ and the initial stress σ^0 to be directed in the X_1 direction. We shall apply algorithm (5.2.13) and solve the problem for two values of σ^u, for the case when the ratio $v_P \Delta t/\Delta X = .5$, say.

First, let $(\sigma^u - \sigma^0)/\Delta\sigma = 1$. The reason for this choice will become clear as we proceed. From Griffith's theory, we can infer that, for a given value of this parameter, there is a critical (minimum) crack size required for crack propagation to occur. If the initially appearing crack were taken just equal to this critical size, crack propagation would proceed very slowly at first before accelerating to values comparable to the wave speeds of the medium. In numerical calculations, it would then be necessary to perform the calculations for a very large number of time steps. To reduce the time required for a crack to reach its terminal fracture speed, we shall take the initial crack to be larger than the critical size. The parameter $(\sigma^u - \sigma^0)/\Delta\sigma$ has often been denoted by s or S in the literature. For $S = 1$, an initial crack size of $7\Delta X$ is supercritical. The crack edge position in time is calculated and plotted in Figure 5.8. Since the crack shape is symmetric about the X_1 and X_2 axes, only one quadrant of the crack is plotted, the other quadrants simply being mirror reflections of this quadrant about the coordinate axes. The initial crack is indicated by the stippled region and the normalized time $T = v_{\mathrm{P}}t/\Delta X$ is marked next to the steplike lines, which indicate the crack edge position at time T. The magnitude and direction of the fracture velocity along the crack front over the last thirty time steps are also indicated on the figure. At the time the calculations were terminated, the crack half-dimension was $30\,\Delta X$ in the X_1 direction and $18\,\Delta X$ in the X_2 direction. No change in the crack edge speed was perceived during the last thirty time steps within the resolution of this numerical scheme. The figure shows that, for the chosen value of S, the terminal crack speed in the X_1 direction (the purely inplane or Mode II direction of fracture) is v_{P} whereas that in the X_2 direction (pure antiplane or Mode III fracture direction) is $0.57v_{\mathrm{P}} \approx v_{\mathrm{S}}$ (for a Poisson solid and within the numerical resolution). Similar results are obtained for $S < 1$ as well as for finer grids, with the value of S being adjusted, of course, for the finer grids (increased in this case) according to our discussion at the beginning of this section. These results agree qualitatively with numerical results from two-dimensional problems, which show that, for cracks propagating on planes with critical stress close to the initial stress, the terminal velocities in the purely inplane and the purely antiplane directions are v_{P} and v_{S}, respectively (Hamano, 1974; Andrews, 1976a, b, 1985; Das and Aki, 1977a; Virieux and Madariaga, 1982; Okubo, 1986). Thus, an initially equidimensional crack will become elongated in the direction of the applied stress on such fault planes.

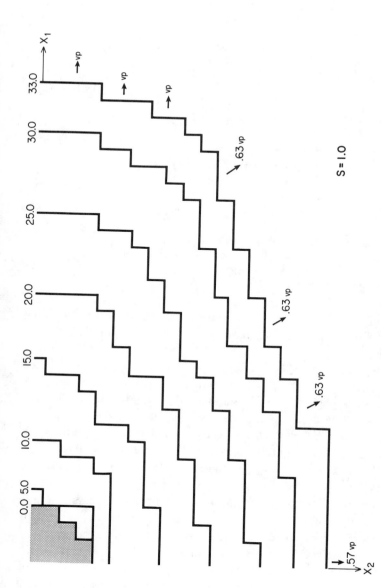

5.8 One quadrant of crack position contours as a function of normalized time, for $S = 1.0$. The normalized time corresponding to each contour is marked along the X_1 axis. The stippled region indicates the initial crack. The crack edge velocity over the last thirty time steps is marked along the last crack edge shown. For this calculation, $v_P \, \Delta t / \Delta X = .5$. (*From Das, 1981.* © *Roy. Astr. Soc.*)

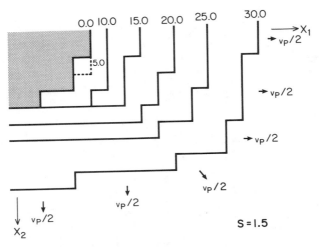

5.9   Same as Figure 5.8 but for $S = 1.5$. (*From Das, 1981.* © *Roy. Astr. Soc.*)

As we go to higher values of S, a remarkable change in these terminal velocities is encountered. Let us consider the particular case when $S = 1.5$. Physically, the fault plane in this case has a higher σ^u or a lower σ^0 or a lower $\Delta\sigma$ than in the previous case. The initially appearing crack is taken to have a diameter of $9\Delta X$ to make it supercritical. Figure 5.9 shows the crack edge position in normalized time units T. At the time calculations are terminated, the crack half-lengths in the X_1 and X_2 directions were $16\Delta X$ and $11\Delta X$. The terminal fracture speed in the purely inplane (X_1) and purely antiplane (X_2) directions were the same and equal to $.5v_p$. For values of S higher than 1.5, similar values of the terminal velocity were found, though, of course, the time required to reach this velocity increased with S. Thus, a transition occurs in crack behavior going from lower to higher values of S. The results agree with two-dimensional results obtained using various numerical schemes. Such a transition was also shown to exist by Burridge (1973) based on analytic considerations. The exact value of S at which this transition occurs depends on the numerical method, the discretization of the problem, and so on and is thus not of general interest. What is important is that there is no physical restriction for inplane shear cracks to propagate at speeds exceeding the Rayleigh wave speed of the medium in nonideally brittle bodies, as we mentioned in Chapter 2.

A possible explanation of this transition is as follows. Higher values of S correspond to higher σ^u and consequently to a slower increase in the

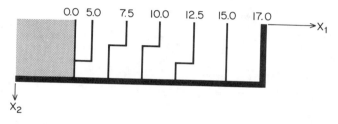

5.10 One quadrant of the crack edge position for an instantaneously appearing crack (stippled) that grows spontaneously to fill up a rectangular area of half-sides $3.5 \Delta X$ by $14.5 \Delta X$. The parameter $S = 1$ and $v_P \Delta t / \Delta X = .5$ for this case. (*From Das, 1981.* © *Roy. Astr. Soc.*)

crack edge velocity. In contrast, the transition from σ^u to the final stress $\sigma^0 - \Delta\sigma$ occurs within a grid. So this numerical model corresponds to a nonideally brittle body with the process zone size less than or equal to the grid size. The case of the ideally brittle body is obtained when the process zone size is negligible compared with other dimensions in a problem. These dimensions in our case are the crack size and the wave front radii at a given time step. With advancing time the grid sizes become smaller and smaller compared with these scales, and the model approaches the limiting case of the ideally brittle body for which the crack speed is limited by v_S in the antiplane and by v_R in the inplane directions. Consequently, the slower the increase in the crack speed with time, the harder it becomes to cross this speed limit of the ideally brittle model.

Spontaneous rectangular shear crack: We shall consider an instantaneously appearing square crack of half-sides $3.5 \Delta X$ that grows spontaneously on a plane under the critical stress level criterion. The final crack shape is constrained to be a rectangle of half-sides $3.5 \Delta X$ by $14.5 \Delta X$ by making $S = 1$ within this rectangle and very large outside it. The propagation here is mainly in the inplane shear mode and the fracture front is shown in Figure 5.10. The slip on the crack is stopped by simply "freezing" the slip at each point on it when the slip attempts to reverse sign, that is, when the slip velocity there becomes zero and reverses direction. In a three-dimensional problem such as this, the slip velocity has two components on the crack plane, and it is quite possible for the slip to reverse direction without passing through zero – for example, if there were some twisting of the crack plane. In this case, this situation did not arise. However, in cases where it does, a simple

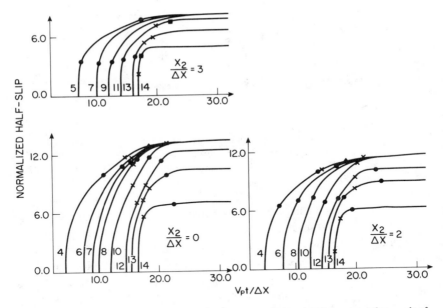

5.11   Normalized half-slip versus normalized time for the rectangular crack of Figure 5.10, plotted for various points on the crack. The X_2 coordinate is shown for each set of curves and the X_1 coordinate is written next to each curve. The half-slip is normalized by $(\Delta X \Delta\sigma)/3\mu$. The dots, crosses, and triangles are explained in the text. (*From Das, 1981. © Roy. Astr. Soc.*)

criterion for stopping slip on the crack will not be applicable, and the more rigorous implementation of friction on a crack that was given in Section 5.5 must then be used.

The calculated half-slip $a_1/2$ as a function of time on several points on the crack is shown in Figure 5.11. The a_2 component of slip is found to be virtually zero. The theoretically expected arrival times of the diffracted shear waves from the lower ($X_2/\Delta\sigma = 3.5$ for all X_1) and upper ($X_2/\Delta X = -3.5$ for all X_1) edges of the crack for points in the quadrant $X_1 > 0$, $X_2 > 0$ are marked in the figure by solid circles, the first circle on each curve representing the arrival from the lower crack edge. At some points on the crack, the expected arrival times of the P and S diffracted waves from the end $X_1/\Delta X = 14.5$ (for all X_2) are denoted by crosses. When only one cross is marked, it corresponds to the P phase. The expected arrival time of the diffracted P wave from the end $X_1/\Delta X = -14.5$ (for all X_2) is denoted by a solid triangle for a few points. To reduce confusion, the expected arrival time of every one of

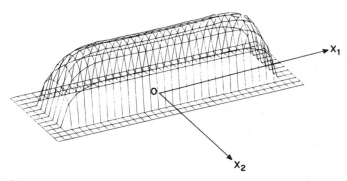

5.12   Perspective plot of final slip on the rectangular crack of Figure 5.10. (*From Das, 1981. © Roy. Astr. Soc.*)

these phases is not plotted on every curve, but the expected arrival time of the first two arrivals among these is always marked. An examination of the slip on the crack relative to these expected arrival times shows that the slip appears to slow down when the first of either the diffracted S wave from the upper or lower crack edge or the diffracted P wave from the crack ends arrives at an interior point. The slip stops soon after the second of these phases arrives at a point on the crack. This can be understood if it is remembered that the slip in this problem is essentially in the X_1 direction and that the S waves from the upper and lower crack edges and P waves from the ends have particle motions in the X_1 direction, whereas the particle motion of the P wave from the upper and lower crack edges is in the X_2 direction. Thus, both the rise time (time for slip to reach its maximum value from zero) and final slip are functions of position on the crack. Over most of the crack region except for the regions close to the ends, the slip stops after the shear waves diffracted by the upper and lower edges of the crack have returned to interior points. This means that the magnitude of the slip on the crack is controlled mainly by the *width* of the crack in such a case. It must be noted that some other mode of stopping the slip may lead to some other result in this respect. An example of the slip being controlled by the crack length is given in Scholz (1982) and Das (1982).

Finally, the plot of the final slip on the crack after all motion has ceased is shown in Figure 5.12. The figure shows that the final slip on a rectangular crack with constant stress drop $\Delta\sigma$ is a function of position on the crack. Since analytic solutions do not exist for static rectangular cracks, numerical results of this type are the only way to estimate the constant c in equation (4.3.3), which relates the average stress drop to the fault dimensions (fault width for the rectangular crack studied here). For this case, in which the length-to-width ratio of the rectangle is ~ 4 : 1, the constant c is found to lie between 0.6 and 0.8, depending on the method of averaging the slip in the discrete solution.

Spontaneous extension of a preexisting circular crack on an infinite plane of constant strength: In the last two examples, we assumed that the initial crack appeared instantaneously and then started extending spontaneously. A more realistic situation is one in which a crack exists at the initial equilibrium state, as we mentioned at the beginning of this section. In this case, the initial stress on the fault plane is not uniform and its deviation from the uniform applied stress is given by the stress distribution around the initial crack. Various ways of determining this initial stress for input into a numerical scheme were also discussed earlier in this section. For this case, let us determine the initial stress numerically by using the boundary–integral method.

Let us consider the problem of a shear crack that is initially of diameter $9 \Delta X$. Let this initial crack be denoted by $S(0)$. The direction of the initial applied stress σ_α^0 is assumed to be constant. For an initially circular crack, we may assume this stress to be applied in the X_1 direction without loss of generality. Let the constant prescribed stress drop $\Delta\sigma$ on $S(0)$ be in the X_1 direction. We shall first solve this problem for the case when the crack edge remains stationary to determine the initial stress. Let $S_{(0)}$ be the discrete form of $S(0)$ and $\overline{S}_{(0)}$ its complement. Then the discrete problem we must solve is

$$\tau_{1ijk} = \Delta\sigma, \ \tau_{2ijk} = 0 \quad \text{for} \quad S_{(0)} : (i\Delta X)^2 + (j\Delta X)^2 \le (4.5\Delta X)^2$$

$$a_{\alpha ijk} = 0 \quad \text{for} \quad \overline{S}_{(0)} : (i\Delta X)^2 + (j\Delta X)^2 > (4.5\Delta X)^2$$

On the basis of the discussion in Section 5.4, we shall allow backslip on the fault to obtain the best static solution. Using equation (5.2.13), calculations are carried out until no further change occurs in the values of the stress on the crack plane and the final solution has been reached. Figure 5.13 shows the numerically obtained stress distribution σ_1^0 along the X_1 and X_2 axes, σ_2^0 being virtually zero as expected from the analytic solution to this problem (Eshelby, 1957; Kostrov and Das, 1984). The ratio of the stress intensity factors at the crack edge in the purely inplane and purely antiplane directions is expected to be $\frac{4}{3}$ for a Poisson solid. The figure shows that the average stresses at the grid

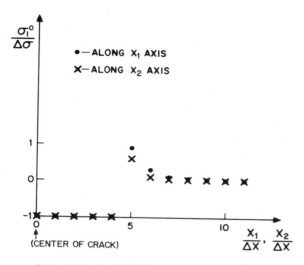

5.13   Normalized initial stress distribution σ_1^0 on a preexisting circular crack along the X_1 and X_2 axes.

adjacent to the crack edge on the X_1 and X_2 axes are in this ratio. This average stress varies smoothly along the crack edge from the end of one axis to the other. The numerically calculated stresses also show the inverse square-root behavior near the crack edge. As we mentioned before, discretization of the analytic solution not only is nontrivial but depends strongly on the method of discretization. What is most important is that the details of the dynamic fracture process calculations for which this initial stress is an input depend very strongly on σ_α^0.

We may now use the σ_α^0 calculated in this way and solve the following dynamic problem:

$$\tau_{\alpha i\,jk} = 0 \quad \text{on} \quad S_{(0)}$$

$$\tau_{1i\,jk} = \sigma_1^0 - \Delta\sigma; \quad \tau_{2i\,jk} = 0 \quad \text{on} \quad \left(S_{(k)} - S_{(0)}\right)$$

$$a_{\alpha i\,jk} = 0 \quad \text{on} \quad \overline{S}_{(k)}$$

Dynamic fracture is initiated by artificial release of one of the two grid points along the crack edge with the highest stress concentration that lie along the X_1 axis for this problem. Figure 5.14 shows the fractured regions of the crack as a function of normalized time for the case when $S = .75$ for the region outside $S_{(0)}$. The fracture process is symmetric about the X_1 axis. The fracture commences propagating from the initially released area at $X_1 = 5.5\,\Delta X$, $X_2 = 0$ along the perimeter of the

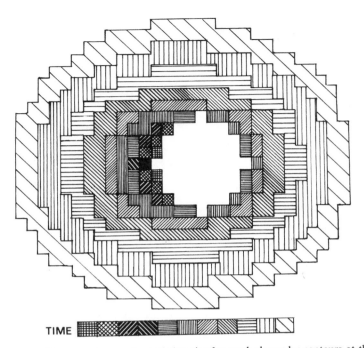

TIME

5.14 Fracture process of preexisting circular crack shown by contours at the fixed normalized time intervals of $v_P t/r_0 \sim 1.4$. The initial crack is unstippled, and the grid that is released to initiate dynamic fracture is shaded black.

initial crack; that is, it follows the stress concentration in the vicinity of this area. At the same time the disturbances generated by the fracturing start spreading out over the $X_3 = 0$ plane and at the normalized time $v_P t/r_0 = 1.7$, where r_0 is the initial crack radius, initiate fracturing close to the point that is diametrically opposite to the fracture initiation point. This is the time of arrival of the S wave at this point from the initially fractured point. A second area of fracture then starts spreading out from this point, the propagation still occurring along the edge of the initial crack. Until time $v_P t/r_0 = 5.5$, fracture propagation occurs mainly in the inplane and mixed modes. At this time the fracture finally commences in the purely antiplane mode as well – that is, the stress concentration in the purely antiplane direction finally reaches the critical stress level required for fracture. The calculations are carried out until $v_P t/r_0 = 15$. By time $v_P t/r_0 = 8.8$ the fracture front has accelerated to its terminal velocity in all directions, this velocity being $v_P/2$, that is, at about the Rayleigh wave speed of the medium, and the crack front is plotted only

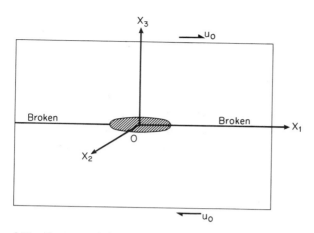

5.15   Geometry of the asperity problem.

up to this time in the figure. From time $v_p t/r_0$ greater than about 10, the difference in its dimension in the X_1 and X_2 directions becomes negligible compared with its actual dimensions.

A circular asperity on an infinite fault plane: This problem was mentioned in Section 4.6, and its far-field radiation patterns were discussed. Here we shall consider the spontaneous fracture of the asperity. This problem is of importance in seismology since some large earthquakes as well as some aftershocks are believed to be due to the presence of such asperities, as we shall see in the next chapter. The fracture of a single asperity on an infinite fault is an oversimplified problem. In the next chapter, we shall consider the radiation due to the fracture of such an asperity but surrounded by a finite fractured area. The present problem is of interest, however, for studying the details of the fracture process of such an asperity, and it is only this case for which the radiation patterns given by (4.6.3) and (4.6.4) are relevant. The results for both the fracture process and the radiation patterns are valid, even in the general case, up to the time when the effects of other asperities on the fault or the main crack edge are felt at the asperity under consideration. In a dynamic problem, this is the time when waves diffracted from the edges of other asperities or the main fault edge return to the asperity under study and modify the solution there.

Let us consider the three-dimensional problem of a preexisting circular asperity of radius r_0 on an infinite plane $X_3 = 0$. At infinity, the two half-spaces are shifted by an amount u_0, say, as shown in Figure 5.15.

Let the initial asperity area (unbroken) be denoted by $S(0)$ and its complement by $\bar{S}(0)$. As before, we shall use σ^u to denote either the fracture strength σ^Y or the frictional strength σ^{stat} on $S(0)$. On $\bar{S}(0)$, the frictional traction is assumed constant. Let us also assume that the displacement of the two half-spaces at infinity occurred quasi-statically and the broken region of the fault plane slipped quasi-statically in response and resulted in concentrating stresses along the boundary of the asperity. The necessary static shear traction components on $X_3 = 0$ can be obtained for a Poisson solid from the well-known solution of Mindlin (1949) as

$$\left. \begin{aligned} \sigma_{13}^{\text{stat}}(\mathbf{r}) &= 16\mu u_0 \big/ \left(7\pi\sqrt{r_0^2 - r^2}\,\right) \\ \sigma_{23}^{\text{stat}}(\mathbf{r}) &= 0 \end{aligned} \right\} \quad \text{on} \quad S(0) : r \le r_0$$

$$\sigma_{i3}^{\text{stat}}(\mathbf{r}) = 0 \qquad\qquad\qquad \text{on} \quad \bar{S}(0) : r > r_0$$

where $\mathbf{r} = (X_1, X_2)$. An obvious property of $\sigma_{13}^{\text{stat}}$ from the above expression is that its value at the center of the asperity is half its average value over the entire asperity.

Next we must discretize the analytic form of $\sigma_{13}^{\text{stat}}$ for use in the numerical algorithm. We may average $\sigma_{13}^{\text{stat}}$ by averaging over grid areas, but as mentioned earlier in this section the method of this discretization is crucial because it must conserve the important properties of the solution. In this example, the traction distribution on the asperity is radially symmetric, but a straightforward discretization of the analytic initial stress distribution over Cartesian grids centered at $(i\,\Delta X, j\,\Delta X)$ destroys this property along the bounding edge of the asperity. The radial symmetry can be preserved by averaging the static traction distribution over polar grids having areas $(\Delta X)^2$ and centered at $(i\,\Delta X, j\,\Delta X)$. The normalized averaged values of $\sigma_{13}^{\text{stat}}$ across any diameter of the asperity is shown in Figure 5.16, the normalization factor being $(16\mu u_0)/7\pi$, for the case when $r_0 = 5.5\,\Delta X$. (There are ninety-seven grids on the asperity in this case.) The expected inverse square-root behavior of the tractions near the edge is seen to be well preserved. The property that the value of $\sigma_{13}^{\text{stat}}$ at the center is half its average value is found to be conserved automatically. It must be reiterated here that some other method of discretization may lead to a different fracturing process. If, however, the most important property, that is, the radial symmetry of the initial stress, is preserved, then the most important features of the resulting fracture process will be the same, although, obviously, the details will differ. The earlier discussions regarding the grid size and its

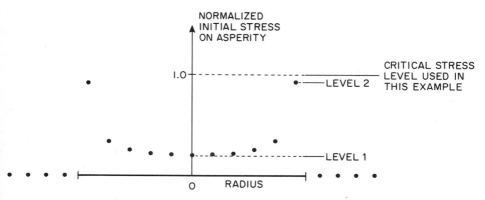

5.16 Discretized initial stress (i.e., negative of the stress drop) distribution, shown along a diameter of the circular asperity. (*From Das and Kostrov, 1983.* © *Am. Geophys. Union.*)

effects on the dynamic solutions must also be kept in mind when considering the results that follow.

This problem is especially suited for algorithm (5.2.9). In particular, if there is no backslip on the fault (and this is the *only* assumption we make regarding the nature of the solution), there is no traction change on $\overline{S}(0)$ during fracture so that the fracture process calculations require that the summations in (5.2.9) extend only over the initially unbroken asperity $S(0)$. Furthermore, the calculations need be performed only over the as yet unbroken part of the asperity $S(t)$, which *decreases* as time increases. Also, once the radius of the Rayleigh cone (see Appendix 1) exceeds the initial asperity diameter $2r_0$, further calculations need be carried out over only a fixed number of time steps. We clarify these statements using Figure 5.17, which shows a cross section through the region of integration in the (X_1, X_2, t) space, the unbroken region of the asperity being indicated by stippling. To determine the stresses at representative points A and B, it is necessary to carry out the convolutions in the region of intersection of the P-wave cone (marked P) having its vertex at A or B, with the cylindrical region of radius r_0. Since in three-dimensional problems the Green function is zero once the Rayleigh wave has passed (the Rayleigh cone is marked R in the figure), the regions over which the convolutions actually have to be carried out are even smaller and are indicated by the hatched regions. Once the radius of the Rayleigh cone exceeds the initial asperity diameter, the figure shows that there is no contribution to the convolution from lower time levels (illustrated for point B). This also implies that to solve this

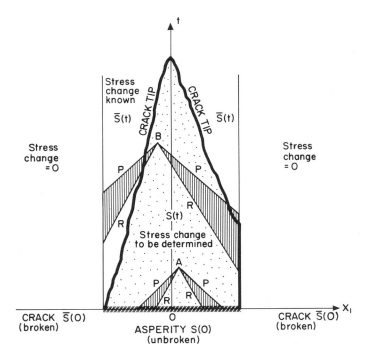

5.17   Section through the volume of integration in the (X_1, X_2, t) space for the asperity problem. (*From Das and Kostrov, 1983.* © *Am. Geophys. Union.*)

problem the Green function need be determined only for a fixed time interval related to the initial asperity size and not up to the time level for which the solution is desired, thereby resulting in further economy in computation.

The spontaneous fracture process of the asperity is to be determined for some given critical stress level. If this level were chosen below level 1 of Figure 5.16, the whole asperity would break at once. If this critical stress level were between levels 1 and 2, all grids along the circumference would break immediately and dynamic fracturing of the remainder of the asperity would occur. The shape of the region that would break immediately would be an annulus, and its width would depend on the particular value of σ^u that was chosen. If the critical stress level were above level 2, the asperity would remain unbroken. If, however, a few points on the asperity were relaxed, the resulting additional stress concentrations might (or might not) cause dynamic fracture to commence. We shall consider the last case in which dynamic fracture does

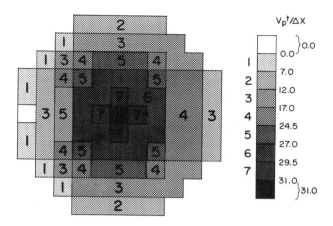

5.18 Fractured areas as a function of normalized times. The key shows the different areas fractured during the time intervals marked. (*From Das and Kostrov, 1983.* © *Am. Geophys. Union.*)

occur. The critical stress level is assumed constant over the asperity in this example. Different distributions of σ^u over the asperity can be easily considered, if so desired. To initiate dynamic fracture, one or more of the most highly stressed grids along the edge of the asperity may be released. Let us release the grid closest to the asperity edge on the X_1 axis, say. The slip on the asperity is found to be mainly in the X_1 direction, the a_2 component of slip being found to be virtually zero, and the traction perturbation component $\tau_{23} \approx 0$.

The fracture process for this case is shown in Figure 5.18. The sequence of fracturing is keyed to the normalized times $v_p t / \Delta X$ shown in the figure by different levels of stippling. The fracture propagation is symmetric about the X_2 axis, as is expected for this case. The fracture propagates along the edge of the asperity in two directions and completely encircles the asperity. The fracture then propagates inward from all directions and finally fractures the central grid. Such a fracture process was termed the "double encircling pincer" process by Das and Kostrov (1983), in analogy with the term for the classic military maneuver. To characterize the rate of fracturing, one can divide the total fracture time by the asperity diameter, which in this case gives $.35v_p$. This value is quite unrelated to the fracture speed locally on the asperity, which is seen in Figure 5.18 to be higher along the asperity edge. A more useful measure of the time history of fracture in such a case may be the number of grid elements fractured at a given time.

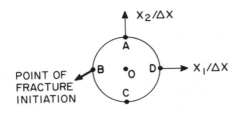

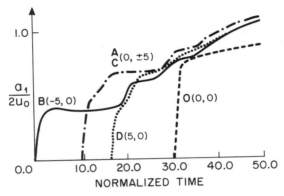

5.19 Normalized half-slip versus normalized time for points *A*, *B*, *C*, and *D* on the asperity. (*From Das and Kostrov, 1983.* © *Am. Geophys. Union.*)

The normalized half-slip $a_1/2u_0$ as a function of time is shown for several points on the asperity in Figure 5.19, the location of these points being indicated in the upper part of this figure. The rise time is found to be shortest for points near the center (which fracture later) and longest for points along the edge (which fracture earlier).

By the release of grids at other points along the circumference of the asperity, an essentially similar fracturing process was obtained (Das and Kostrov, 1983).

Failure of an elliptical asperity on an infinite fault: This is the same problem as that discussed in the last subsection, but now the asperity has an elliptical shape. This problem is more complicated than the previous one for two obvious reasons. First, there is no radial symmetry in the initial traction distribution. Second, the asperity may be oriented at some arbitrary direction to the initial shift, so that the initial traction on the asperity is not generally parallel to a coordinate axis.

Let the initial asperity area $S(0)$ be given by

$$\frac{X_1^2}{a^2} + \frac{X_2^2}{b^2} \le 1; \qquad X_3 = 0$$

and let the applied shift $\pm u_0$ at infinity make an angle θ to the X_1 axis so that its components in the X_1 and X_2 direction are

$$a_1^{\text{stat}} = u_0 \cos \theta$$

$$a_2^{\text{stat}} = u_0 \sin \theta$$

The initial traction distribution on the asperity was given by Mindlin (1949) and by Das and Kostrov (1985) as

$$\sigma_\alpha^0(X_1, X_2, 0) = \frac{P_\alpha}{\sqrt{1 - X_1^2/a^2 - X_2^2/b^2}}$$

where P_α depends on a, b, and θ. Its main properties are that its magnitude rises from one-half of the average at the center to infinity at the edge and its direction is constant over the asperity but makes some angle, say ϕ, with the X_1 axis, where $\phi > \theta$ in general. If $\theta = 0$ or $\pi/2$, then $\phi = 0$ or $\pi/2$, respectively. The stress-intensity factor is always highest at the end of the major axis of the ellipse and decreases smoothly to its value at the end of the minor axis. Complete expressions for P_α are given in Das and Kostrov (1985). The initial traction distribution is discretized simply by sampling the initial tractions at grid points on $S(0)$, except for the grids located near the asperity boundary. For grid points closer to the edge than half a grid length, we assign the value of the traction calculated for the point at a distance of half a grid length from the edge that has the same normal to the edge as the grid point under consideration. This approach removes the artificial oscillations that would be introduced by discretizing the tractions over Cartesian grids and preserves all the important properties of the analytic solution.

We shall assume that σ^u is constant over $S(0)$. σ^u cannot be lower than the level of the initial (discrete) traction at the center [lowest stressed point on $S(0)$], for then the asperity could not exist at all. If σ^u is higher than the initial (discrete) traction closest to the edge on the major axis (highest stressed point), then the contact area cannot fracture. If σ^u lies anywhere between these highest and lowest levels, some portion of the asperity will fail and fracture may then continue dynamically or stop, depending on σ^u. Since it is obvious that the stress concentration on $S(0)$ cannot exceed the level σ^u for the asperity to exist

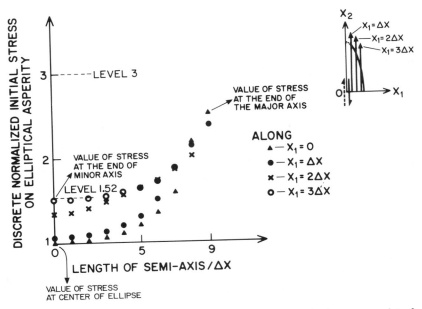

5.20   Discretized initial stress (negative of the stress drop) at every point of elliptical asperity of aspect ratio 3:1. (*From Das and Kostrov, 1985.* © *Roy. Astr. Soc.*)

and since this concentration is directly proportional to the curvature of the asperity perimeter, this suggests that the existence of asperities with sharp corners is disfavored.

For numerical illustrations, we consider the cases when $a = 3 \Delta X$, $b = 9 \Delta X$, and $a = 2 \Delta X$, $b = 10 \Delta X$, for $\theta = 0°$, $45°$, and $90°$. The problem is solved using algorithm (5.2.9).

Figure 5.20 shows the initial traction level (normalized by P_α) at every point of the asperity of aspect ratio $3:1$, and Figure 5.21 shows the fracture process for the case when $\sigma^u/P_\alpha = 3.0$ over the asperity and for $\theta = 90°$ (case A), $\theta = 0°$ (case B), and $\theta = 45°$ (case C). This level of σ^u/P_α is larger than the normalized stress level of the highest stressed point on the asperity (the latter being at level 2.6 in Figure 5.20). Since P_α depends on θ, the actual value of the critical stress is different for the cases with various θ's considered but is three times the value of the initial stress at the center for all the cases. It is found that dynamic fracture could be initiated only by relaxing a grid point at or close to the end of the major axis for all the cases. The point of fracture initiation is indicated by an asterisk in Figure 5.21.

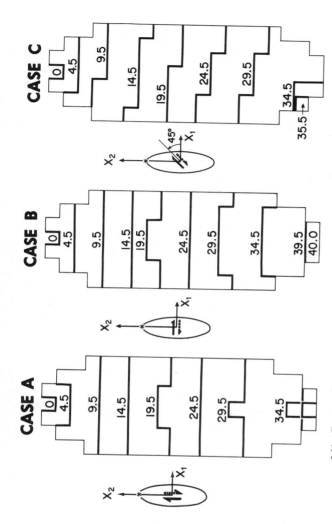

5.21 Fracture front position at normalized time for the elliptical asperity in cases A, B, and C. The direction of force applied to the asperity is shown by thick arrows in the inset to the left of each case, and the asterisk denotes the point of fracture initiation. (*Adapted from Das and Kostrov, 1985.*)

For case A ($\theta = 90°$), the only nonzero component of the initial traction on the asperity is σ_2^0 and results in the traction perturbation component τ_1 being antisymmetric in X_2 and τ_2 being symmetric in X_2. Hence, the fracture front is symmetric about the X_2 axis. It is found that in this case the fracture propagated from one end of the major axis to the other in almost a straight line at an average speed of $\sim v_P/2$. The slip on the fault is found to be essentially in the X_2 direction, and the fracture thus propagates mainly in the inplane mode. So we see that the elliptical asperity does not first fracture along the edge where the initial stress is concentrated. The fracturing process is a result of the interaction between the preexisting initial static stress concentration along its edge and the dynamic stress concentration at the fracture front. The marked difference in the fracturing process for a circular and an elliptical asperity implies that in the circular case the dominant stress concentration was the initial one, whereas for the elliptical case it is the dynamic one. In fact, occasionally the fracture front for the asperity does propagate slightly faster along the edge, but this is not a particularly significant effect.

For case B ($\theta = 0°$), the only nonzero component of the initial traction on the asperity is σ_1^0, so that now τ_1 is symmetric and τ_2 is antisymmetric in X_2 and the fracture front is symmetric about the X_2 axis. The fracture is again initiated at the end of the major axis and found to propagate as a practically straight line from one end of the major axis to the other at an average speed of $\sim .45 v_P$. The slip on the asperity is essentially in the X_2 direction, so that the mode of fracture propagation is mainly antiplane.

For case C ($\theta = 45°$), both components σ_1^0 and σ_2^0 of the initial traction are nonvanishing and are found to be 68 and 73 percent, respectively, of the total traction. The solution to the problem is the sum of two solutions, one with σ_1^0 as the initial stress leading to τ_1 being symmetric and τ_2 being antisymmetric about the X_2 axis and the other with σ_2^0 as the initial stress leading to τ_1 being antisymmetric and τ_2 being symmetric about X_2. The total solution therefore has no line of symmetry on the fault plane. The fracture is initiated at one end of the major axis and is found to propagate faster along the edge in the $X_1 > 0$ region than along the $X_1 < 0$ edge. This results in a rotation of the fracture front as it propagates from one end of the major axis to the other. This reflects the fact that, for the σ^u/P_α chosen, the inplane mode propagates faster than the antiplane mode.

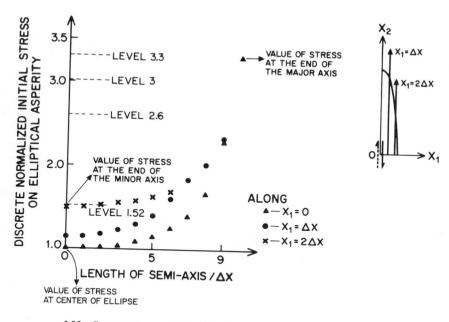

5.22 Same as Figure 5.20 but for the elliptical asperity of aspect ratio 5 : 1. (*From Das and Kostrov, 1985.* © *Roy. Astr. Soc.*)

Thus, it is seen that in spite of the increased complexity of the elliptical asperity problem over the circular problem, the fracture process actually looks simpler. This means that more and more complex asperity geometries and initial traction distributions do not necessarily lead to more and more complex fracture processes!

Figure 5.22 shows the normalized initial traction at every point of the elliptical asperity of aspect ratio 5 : 1. Figure 5.23 shows the fracture process for the following cases:

Case D: $\sigma^u/P_\alpha = 3.0$, $\theta = 90°$
Case E: $\sigma^u/P_\alpha = 2.6$, $\theta = 0°$ (in this case for higher values of σ^u/P_α, dynamic fracture could not be initiated by releasing one or even a few points at the end of the major axis)
Case F: $\sigma^u/P_\alpha = 3.3$, $\theta = 45°$ (the case $\sigma^u/P_\alpha = 3.0$ for $\theta = 45°$ was also studied and the fracture process was qualitatively similar to case D and is not plotted here).

These three levels of σ^u/P_α are indicated in Figure 5.22. In case D, the

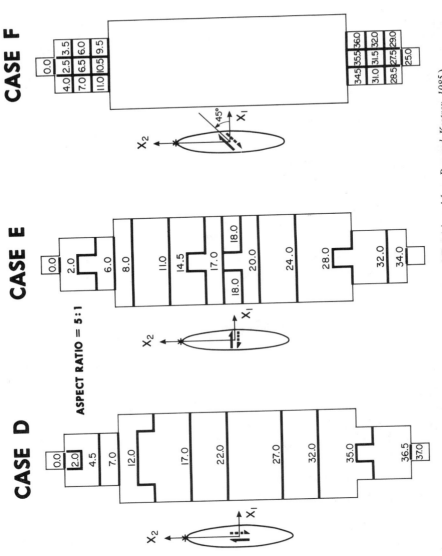

5.23 Same as Figure 5.21 but for cases D, E, and F. (*Adapted from Das and Kostrov, 1985.*)

initial traction is applied along the major axis. Since the level of the normalized σ^u used in this case is lower than the initial stress at the two ends of the major axis, the fracture would naturally commence at both ends of the major axis and propagate in toward the center. In a realistic setting, it is unlikely that both ends of the major axis would attain the critical stress level simultaneously, so fracture is allowed to nucleate only one end of the major axis, the other being prevented from breaking initially by artificially decreasing its initial stress by about 20 percent (or, equivalently, by increasing its critical stress level σ^u). It is found that the fracture propagates from one end of the major axis to the other at about the Rayleigh wave speed of the medium. An essentially similar result is obtained for case E. Finally, in case F, the σ_1^0 and σ_2^2 are found to be 63 and 78 percent of the total traction drop and the fracturing is initiated at the end of the major axis. This highly stressed area fractures, but further fracture of the asperity does not occur until the stress waves from this end reach the other end of the major axis. Some time after the arrival of the compressional waves but before the arrival of the shear waves fracture commences at this second highly stressed region of the asperity and it fails, the remainder of the asperity being left unbroken. These results may of course be modified by a more refined gridding and are presented mainly for the purpose of illustration.

An inplane crack under a linear slip-weakening law: Let us consider a dynamic plane strain shear crack with a linear slip-weakening law. This problem was studied by Andrews (1976a) using a finite difference method and by Andrews (1985) using the relation (5.2.9). We shall base our discussion on the latter work.

A time–distance plot of contours of slip is shown in Figure 5.24 for the case when $S = .8$. The distance and time are normalized, respectively, by L_c and L_c/v_S, where L_c is the half of the critical length of a Griffith crack given by

$$L_c = \frac{2\mu}{\pi}\left(1 - \frac{v_S^2}{v_P^2}\right)\left(1 + \frac{\sigma^u - \sigma^0}{\Delta\sigma}\right)a_0$$

where the critical slip-weakening displacement a_0 (Figure 5.7) is taken as $1.3L_c\,\Delta\sigma/\mu$. The initial crack is taken to be supercritical. The spatial grid length $\Delta X = L_c/5$ and $v_S\,\Delta X/\Delta t = .5$. The crack propagation is bilateral, and only half the crack plane is shown in Figure 5.24, the other half being a mirror reflection about the time axis. The cohesive zone is

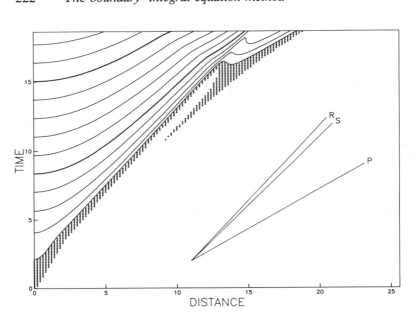

5.24 Time–distance plot of contours of slip due to fracture under a slip-weakening law. The cohesive zone is indicated by squares, the contour interval being a_0, the critical slip-weakening displacement. (*From Andrews, 1985.* © *Seismol. Soc. Am.*)

indicated by small square symbols plotted in each grid element where the slip velocity is nonvanishing and the slip is less than a_0. The fracture is driven at half the Rayleigh wave speed $v_R/2$ until it propagates spontaneously beyond $X_1 = 2L_c$ and rapidly approaches v_R. As the crack lengthens, the cohesive zone shrinks to a width of $3\Delta t$ in time or $1.5\Delta X$ in space. At $X_1 = 9.2L_c$, the leading edge of the cohesive zone jumps to the traction peak at the S wave. Later, the trailing edge of the cohesive zone jumps ahead to $X_1 = 14.4L_c$ and propagates at the speed v_P from there on. A minimum in slip velocity continues to propagate at the speed v_R. Freund (1979) also discussed such a jumping ahead of the crack tip for small values of S, and the resulting primary and secondary crack tips that eventually coalesce.

The fracture speed of such a crack depends on the parameters S and L/L_c, where L is the instantaneous crack half-length. The transition of the fracture speed from sub-Rayleigh to near-P speeds can be seen in Figure 5.25, where the domains of these speeds are shown in the parameter space of S and L_c/L for the four values of S given by $S = .5$, .667, .8, and 1.0. Each fracture was started at the speed $v_R/2$, and the

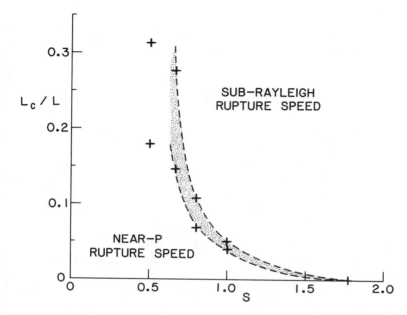

5.25   Domains in parameter space ($L_c/L - S$) of fracture speed of plane-strain shear cracks propagating under a linear slip-weakening law. The shaded region is the zone of transition of the fracture speeds from sub-Rayleigh to near-P speeds. (*From Andrews, 1985. © Seismol. Soc. Am.*)

beginning and end of the transition in speed for each case is indicated by crosses. The continuous lines are drawn through these crosses and extrapolated to higher values of S to find the value where the fracture speed never exceeds v_R. This value is at $S \approx 1.77$ and agrees with the analytic calculations of Burridge (1973) for cracks with friction but without cohesion. The numerically calculated transition zone in Figure 5.25 obtained by Andrews (1976a) and by Andrews (1985) agree within 25 percent. Similar results were found by Okubo (1986) using a more complicated slip-weakening behavior.

These examples indicate that inhomogeneous stress, strength, and friction on a fault can be easily incorporated into the numerical approach.

5.7   General formulation of the boundary conditions on the fault

The integral representations developed at the beginning of this chapter when taken with the constitutive relations on the fault comprise a system of integral equations for slip and traction perturbation on the crack plane. In the last few sections, several examples of possible constitutive relations were given. In general, the constitutive relations may reflect different aspects of the fault behavior such as fracture, friction, joint dilatancy, and static fatigue and would be some functionals relating traction- and slip-time histories, namely,

$$S_1\left(\mathbf{a}|_0^t, \sigma|_0^t\right) = 0; \qquad S_2\left(\mathbf{a}|_0^t, \sigma|_0^t\right) = 0 \qquad (5.7.1)$$

After discretization, these relations would take the form

$$S_\alpha\left(\mathbf{a}_{ijk}, \sigma_{ijk}, \theta_{ij}|_0^{(k-1)}\right) = 0, \qquad \alpha = 1,2, \qquad k = 0,1,\ldots \qquad (5.7.2)$$

where $\theta_{\alpha ij}$ incorporates all the previous slip and traction histories on the fault as well as other possible constitutive parameters.

The above relations are nonlocal in terms of time. In analytic considerations it is usually easier to adopt a simplifying assumption that the relations involve only local (instantaneous) values of slip, slip rate, and traction vectors, which leads to the particular cases discussed in Chapter 1. In such cases the discrete form (5.7.2) will include only dependence on slip values at the previous time step $(k - 1)$. To obtain $\mathbf{a}_{ijk}$ and the traction perturbation τ_{ijk} at a given time step k, one has to solve the set of four simultaneous equations given by (5.2.9) or (5.2.13) and (5.7.2), the latter two being, in general, nonlinear.

In the case of viscous friction or slip-weakening fracture or friction, the set of simultaneous equations (5.7.1) can be solved for $a_{\alpha ijk}$, giving

$$a_{\alpha ijk} = U_\alpha\left(\mathbf{a}_{ij(k-1)}, \tau_{ijk}\right) \qquad (5.7.3)$$

where U_α is some known function. For the case of Coulomb friction, (5.7.1) can be solved for τ_{ijk} to give the form

$$\tau_{\alpha ijk} = T_\alpha\left(\mathbf{a}_{ijk}, \mathbf{a}_{ij(k-1)}\right) \qquad (5.7.4)$$

where T_α is some known function.

6

FAR-FIELD RADIATION FROM NUMERICAL SOURCE MODELS

In this chapter, we describe the method of determining the far-field source function and then consider the pulse shapes and spectra radiated by some specific earthquake source models.

6.1 The initial pulse shape

The expressions for the body wave displacement at a seismic station located at a distance R from the source, R being much larger than the source dimension for an inhomogeneous earth, were given in Chapter 3 [equations (3.3.20), (3.3.22), and (3.3.23)]. This representation is somewhat incomplete, for it does not take into account the filtering effects of the medium and the instrument. In general, the pulse shape $A^P(t', \mathbf{m})$ at a station can be represented as the convolution of the "initial pulse shape" $A^P_{initial}$ with the instrument and medium responses for the P wave, say, as

$$A^P(t', \mathbf{m}) = I(t) * M(t) * A^P_{initial}(t', \mathbf{m})$$

where $I(t)$ and $M(t)$ are, respectively, the instrument and medium response and $\mathbf{m}$ is the ray direction. Formula (3.3.21), which represents, strictly speaking, the source pulse shape A^P_{source}, can be expressed in terms of the slip rate vector for a planar fault as follows:

$$A^P_{initial}(t', \mathbf{m}) = \frac{v_S^2 m_\alpha m_3}{2\pi v_P^3} \int_{\Sigma_1} \dot{a}_\alpha \left(\mathbf{X}, t + \frac{m_\beta X_\beta}{v_P} \right) dS \qquad (6.1.1)$$

where, as before, $X_3 = 0$ is the fault plane and $\dot{a}_\alpha$ is the slip rate vector. If the slip rate *direction* is assumed to be constant over the fault throughout the fracturing process, then this formula can be split into two factors, namely, the radiation pattern $\mathscr{P}^P(\mathbf{m})$ for a concentrated dipole given by

$$\mathscr{P}^{P}(\mathbf{m}) = b_{\alpha} m_{\alpha} m_3 \frac{v_S^2}{2\pi v_P^3}$$

b_{α} being the unit vector in the direction of the slip rate vector, and the "source pulse shape" A_{source}^{P} given by

$$A_{\text{source}}^{P}(t', \mathbf{m}) = \int_{\Sigma_1} \dot{a}\left(\mathbf{X}, t + \frac{m_i X_i}{v_P}\right) dS \qquad (6.1.2)$$

where $\dot{a} = b_{\alpha} \dot{a}_{\alpha}$. Similar relations hold for S waves. This "source pulse shape" depends on the ray direction and wave type. When this dependence can be neglected, it is usually called the "source time function." This neglect is valid only for very long wavelengths compared with the source dimension. In general, when the slip rate direction is variable, the initial pulse can be split into the sum of two terms, each being the product of a radiation pattern and a source pulse. In the examples that follow later in this chapter, we shall limit ourselves to considering only the initial pulse shape and use (6.1.1) for selected directions from the source. This representation is convenient for models with a confined slipping region.

An alternative representation for $A_{\text{initial}}^{P}(t', \mathbf{m})$ can be obtained from (4.6.3) in terms of traction perturbations on the entire fault plane,

$$A_{\text{initial}}^{P}(t', \mathbf{m}) = \frac{v_S |m_3| m_{\alpha}}{\pi \rho v_P^3 \mathscr{F}^{P}(\mathbf{m})} \int_{\Sigma} \tau_{\alpha}\left(\mathbf{X}, t + \frac{m_i X_i}{v_P}\right) dS \qquad (6.1.3)$$

where

$$\mathscr{F}^{P}(\mathbf{m}) = 4\left(\frac{v_S}{v_P}\right)^2 |m_3|(1 - m_3^2)\left[1 - \left(\frac{v_S}{v_P}\right)^2(1 - m_3^2)^{1/2}\right]$$

$$+ \left[1 - 2\left(\frac{v_S}{v_P}\right)^2(1 - m_3^2)\right]^2 \qquad (6.1.4)$$

This representation of the initial pulse shape is more convenient for problems with limited areas of perturbed tractions.

For numerical computation of (6.1.1) or (6.1.2) for crack (dislocation) problems, it is more convenient (and more stable) to first evaluate the time integral of the pulse shape $\int_0^t A_{\text{initial}}^{P} dt$ and then differentiate it numerically to obtain the pulse shape. The advantage of this method is that the final value of the integral is proportional to the static seismic moment. In discrete form, we have

$$\int_0^{k\Delta t} A_{\text{initial}}^{P}(t', \mathbf{m}) \, dt \approx \frac{v_S^2 m_{\alpha} m_3}{2\pi v_P^3} \sum_{i,j \in \Sigma_k} a_{\alpha i j k'} \qquad (6.1.5)$$

where

$$k' = \text{integer}\left[k - \frac{m_1 i + m_2 j}{v_P \Delta t} \Delta X \right]$$

and Σ_k is the discrete crack area. When using (6.1.3), the pulse shape is given in discrete form by

$$A_{\text{initial}}^P (k \, \Delta t, \mathbf{m}) \, dt \approx \frac{v_S^2 m_\alpha |m_3|}{\pi \rho v_P^3 \mathscr{F}^P(\mathbf{m})} \sum_{i, \, j \in \Sigma_k} \tau_{\alpha i j k'} \qquad (6.1.6)$$

where k' was defined above. For the isolated single-asperity problems discussed in the earlier chapters, the final value of the pulse shape is proportional to the total "static force drop" on the asperity.

In the next sections, some specific radiated pulse shapes and their amplitude spectra are discussed.

6.2   Far-field radiation from simple faulting models

A circular crack that propagates at a speed $v_P/2$ and arrests when it reaches some given size was discussed in Section 5.4. We shall determine the pulses radiated to the far field by such a propagating crack. The normalized P-wave pulse shape is the time derivative of the term

$$\sum_{i, \, j \in \Sigma_k} a_{\alpha i j k'}$$

in equation (6.1.5). The P and S pulse shapes and their spectra are shown in Figure 6.1 for directions $\theta = 0°$, $45°$, and $90°$, θ being the angle that the direction to the source makes with the normal to the fault plane. The ray direction $\mathbf{m}$ can be simply expressed in terms of θ as $m_1 = \sin\theta$, $m_2 = 0$, $m_3 = \cos\theta$.

The far-field radiation varies significantly as a function of θ, the most significant difference being between the radiation along the normal to the fault ($\theta = 0°$) and the other directions from the source. Along the normal, the far-field radiation as a function of time comes from concentric circles on the fault, which are centered at the point of crack nucleation and which spread out at a speed $v_P/2$. (In the discrete form, the far-field contribution comes from concentric annular regions on the fault.) As one moves away from the normal, the far-field radiation comes from asymmetric closed curves around the nucleation point. This results in less constructive interference between the waves than that along the normal and thereby lowers the amplitude of the radiated pulse off the

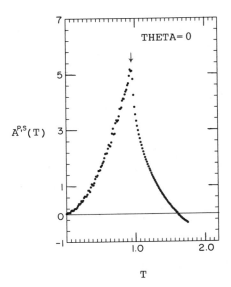

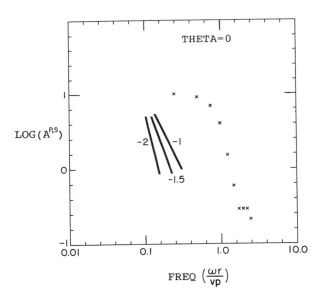

6.1   Normalized far-field P- and S-wave pulse shapes $A^P(T)$ and $A^S(T)$ and the spectra log $A^P(\omega)$ and log $A^S(\omega)$ in the directions $\theta = 0°$, $45°$, and $90°$ from the normal to the crack. The pulse shapes are plotted against normalized time $T = v_P t / \Delta X$, measured from the time of arrival of the first wave at the observer. The spectra are plotted against the normalized frequency ($\omega r / v_P$), r being the final crack radius. The spectra : normalized by their value at zero frequency.

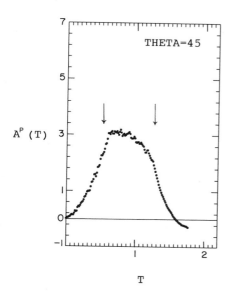

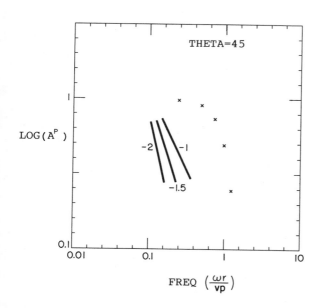

6.1   *(cont.)*

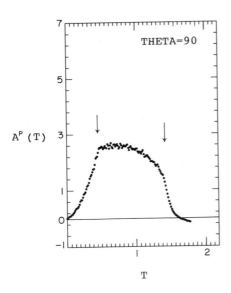

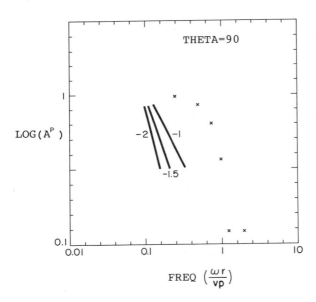

6.1   *(cont.)*

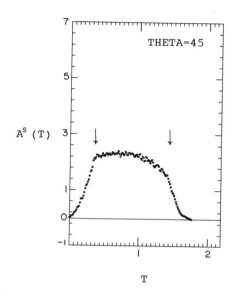

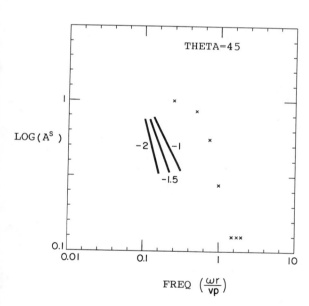

6.1   (*cont.*)

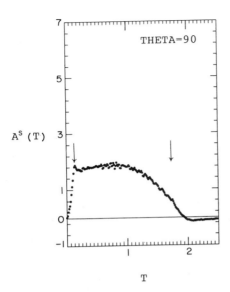

6.1   (*cont.*)

normal. The geometric arrival times of the diffracted phase from the nearest and farthest crack edges is given by

$$t = r\left(\frac{1}{v} \pm \frac{\sin\theta}{c}\right) \tag{6.2.1}$$

where $c = v_P$ or v_S for P or S waves, respectively, and r is the final crack radius, which was taken as $41\,\Delta X$ in this example. These times are indicated by arrows in Figure 6.1, the time axis in these figures being measured from the arrival time of the first wave at the receiver. Clearly, these edge phases control the pulse width. The area under the pulse for any direction for either wave type is the same constant and is proportional to the seismic moment of the earthquake.

The amplitude spectra have three distinct regions. The low-frequency flat portion of it is controlled by the seismic moment, and at these long wavelengths the source looks like a point. Next, an intermediate frequency region in which the amplitude is lower than the flat part of the spectrum follows. The spectral amplitude in this region is controlled by waves of wavelengths related to the duration of motion on the fault (and for constant fracture speed these wavelengths are simply related to the fault dimension). The frequency where this intermediate spectral decay commences is called the "corner frequency" of the source. Figure 6.1 shows that the rate of decay of the spectral amplitude in this range depends on the direction θ to the receiver. Finally, there is a high-frequency portion that falls off even faster due to the effect of interference of waves on the fault of wavelengths smaller than the fault. On the spectra shown in Figure 6.1, lines having falloffs ω^{-1}, $\omega^{-1.5}$, and ω^{-2} are shown to aid the eye in determining the rate of spectral decay. It is seen that at the highest frequencies the spectrum falls off as ω^{-2} for $\theta > 0°$.

The problem described here was first studied by Madariaga (1976). Following him, we next consider the P- and S-wave corner frequencies for different values of θ for cracks propagating at various fracture speeds, and the results are shown in Figure 6.2. The figure shows that for $\theta > 30°$, the P-wave corner frequency is higher than the S-wave corner frequency for all the cases considered. This is a result of the longer duration of the S pulse compared with the P pulse, which is due to the longer delay time between the arrival of the edge phases for the S wave than for the P wave [from (6.2.1)]. This effect is clear in measured corner frequencies (Molnar, Tucker, and Brune, 1973).

For problems without the simple geometry of this problem, the effects discussed here will become more complicated. In particular, the pulse

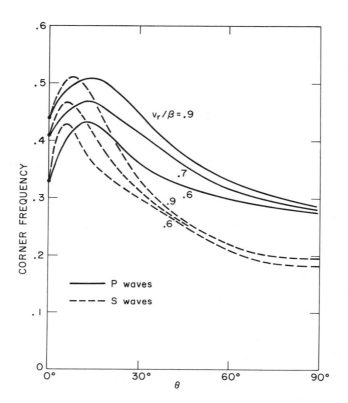

6.2   P- and S-wave corner frequencies looking in different directions from source for cracks propagating at different speeds. (*Adapted from Madariaga, 1976.*)

width, corner frequency, and spectral decay rates in the forward and backward fracture propagation directions will be different. These effects will be clear from the results of the unilateral problem to be illustrated in Section 6.3.

6.3   The heterogeneous faulting process

We saw in the last chapter that the fracture process depends on the initial stress distribution on the fault and on the physical properties of the fault such as its strength and its static and dynamic friction levels. Variations in any one of these parameters over the crack plane would produce variations in the fracture velocity, slip rate, and stress drop distribution over the fault. This heterogeneity would be manifested in the complexity of the radiated pulse shapes. Such observations of "multiple shocks" led to the introduction and subsequent acceptance of models with heterogeneous stress drop and strength over the fault plane. Two idealizations of this situation have been considered in the past decade. In one, known as the "barrier" model, the stress drop on the fractured part of the fault plane is essentially uniform and the critical stress level has large variations. In the other, known as the "asperity" model, the stress drop is highly variable over the fault. Obviously, every conceivable variation and combination of these two extreme cases is plausible in reality. Also, instead of one unique crack edge, there may be multiple crack edges due to the locking of regions behind the main crack edge. The stress drop in this case becomes inhomogeneous not only in space but in time as well. A problem of random variation of stress drop and strength over the fault was studied numerically by Mikumo and Miyatake (1979), though with a somewhat simplified model. The fracture process was found to be quite chaotic, with no clearly distinguishable fracture front. In such cases, a stochastic or a fractal approach may be instrumental.

It is now well known that some aftershocks occur off the main fault plane. Obviously, a complex seismic event may be accompanied by such shocks, occurring during the main earthquake rather than after it. This implies that at least part of the complexity of seismic radiation cannot be assigned to the main fault plane as is assumed in the models mentioned here. This is especially true when one is considering the high-frequency radiation from an earthquake.

In this section, we shall consider only some simple examples of these variations confined to the fault plane and determine the far-field seismic radiation due to fracture propagation on such a plane.

The barrier model

A barrier may be characterized by some measure of its areal extent and some measure of its strength. We may use the parameter S, defined in Section 5.6, to denote the relative strength of the barrier. If the areal extent of the barrier is large, the crack edge propagation will be arrested. But if its areal extent is small compared with the instantaneous crack dimension at the time it is encountered by the crack edge, the crack edge and the barrier will interact in the three different ways, depending on the value of S:

1. If S is small, the barrier will be broken as the crack edge encounters it.

2. If S is very large, the crack edge will propagate around it, leaving behind an unbroken region.

3. If S has some intermediate value, the barrier will not be broken the first time it is encountered by the crack front but will eventually break due to the subsequent concentration of stress on it during the dynamic growth and slip of the surrounding areas.

The presence of such barriers on the fault will introduce diverse slip functions over the fault, which in turn will be seen as complexity in the radiated seismic wave forms and will modify the seismic moment of the earthquake. This problem was first studied two dimensionally by Das and Aki (1977b) and by Das (1985) for the three-dimensional case. Since the two-dimensional results are very complete and are now well known, we shall include only these results here. The three-dimensional calculations show that the two-dimensional results correctly predict the complexity of the far-field waveforms for a fault with barriers.

Let us confine our discussion to the unilateral propagation of an inplane crack. The four cases studied are listed below. The total crack length is taken as $10 \Delta X$ and $S = 0$ in the areas without barriers for all the cases. The latter parameter value means that a critical crack length of zero is needed for dynamic propagation, which makes the calculations very economical. Backslip will not be allowed in all cases so that $a_{\alpha i j k_2'} > a_{\alpha i j k_1'}$, where $k_2' > k_1'$. Hence $A_{\text{initial}}^{\text{P}}(k \Delta t, \mathbf{m})$ is positive for all time and the maximum value of the amplitude spectrum is at zero frequency.

Case P-SV-0: There are no barriers on the fault, this case being included purely for the purpose of comparison. The crack extends at a speed close to v_{P} due to S being chosen as zero. The distribution of S and the resulting slip on the fault are shown in Figure 6.3. The far-field P pulse shape determined from equation (6.1.4) and the corresponding amplitude spectra are given in Figure 6.4.

Case P-SV-1: One strong barrier exists on the fault, and it remains unbroken when the dynamic fracture process on the fault is completed. The distribution of S and the slip on the fault are shown in Figure 6.5, and the far-field radiated field is plotted in Figure 6.6. The spectra for the case P-SV-0 is indicated in the latter figure by dashed lines.

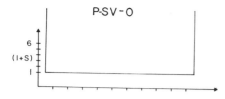

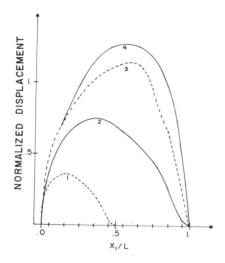

6.3 Distribution of the parameter S and "snapshots" of the distribution of the normalized half-slip $a_1/2$ over the fault length for case P-SV-0. The half-slip is normalized by $L \Delta\sigma/3\mu$, where L is the fault length and the integer next to each curve indicates the time measured in units of $.5L/v_P$. (*From Das and Aki, 1977b. © Am. Geophys. Union.*)

Case P-SV-2: Two unbreakable barriers exist on the fault. Figure 6.7 and Figure 6.8 show the corresponding S, the slip on the fault, and the far-field radiation. The dashed lines again give the P-SV-0 spectra.

Case P-SV-3: The two barriers on the fault, having intermediate S values, do not break at the initial passage of the fracture front but break before the completion of the dynamic fracture and slipping process is completed. The related parameters and results are shown in Figures 6.9 and 6.10.

The major conclusions drawn from this example can be summarized as follows:

1. The smooth fault P-SV-0 and the P-SV-3 fault result in single earthquakes, whereas the heterogeneous faults P-SV-1 and P-SV-2 result in multiple shocks.

2. The time history of slip on the fault and the resulting far-field radiation are most complicated in the case when the initially unbreakable barrier eventually breaks (P-SV-3). In this case the duration of the fracture and slipping process are longer than in the other cases for the same fault length.

3. The final slip on the fault and hence the seismic moment are largest for the smooth crack (P-SV-0) and smallest for the case of the fault with two unbroken barriers (P-SV-2). In the case of the barrier that eventually breaks, the final slip and moment are almost as large as those for the smooth fault. The slip for the fault with two unbreakable barriers has the most uniform value over the fault, whereas the fault with no barriers at the end of the fracture process (P-SV-0 and P-SV-3) shows the largest amount of variation in slip distribution over the fault! This may explain why the uniform dislocation model (Haskell, 1964) has often been able to explain observed overall features of seismograms satisfactorily.

4. Clear directivity effects in the seismic radiation are seen in all cases, these effects being stronger for the fault with unbreakable barriers than for the smooth fault. However, when the barriers eventually break the directivity effect is even weaker than that for the smooth fault.

5. The time domain pulses are more sensitive to the complexity of the fracture process than the spectral shapes. In particular, when the barriers eventually break the pulses show complexity in all directions from the source, but the spectra are not particularly revealing.

6. When the barriers remain unbroken, the spectra at the highest frequencies for which the numerical results are meaningful (this

6.4 Far-field P-wave displacement pulse shape and amplitude spectra for various directions from the fault for case P-SV-0. The angle θ is measured from the normal to the fault. The arrows indicate the arrival of the first diffracted wave when the crack tip stops. For $\theta = 0°$ the P- and S-wave pulse shapes coincide. The amplitude spectra are normalized by their value at zero frequency. (*From Das and Aki, 1977b.* © *Am. Geophys. Union.*)

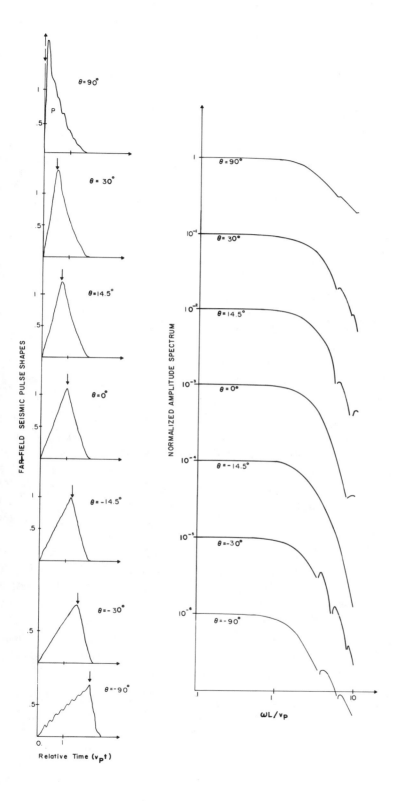

FAR-FIELD SEISMIC PULSE SHAPES

$\theta = 90°$

P

$\theta = 30°$

$\theta = 14.5°$

$\theta = 0°$

$\theta = -14.5°$

$\theta = -30°$

$\theta = -90°$

0.

Relative Time ($v_P t$)

NORMALIZED AMPLITUDE SPECTRUM

$\theta = 90°$

$\theta = 30°$

$\theta = 14.5°$

$\theta = 0°$

$\theta = -14.5°$

$\theta = -30°$

$\theta = -90°$

$\omega L / v_P$

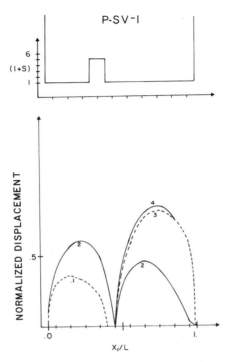

6.5   Same as Figure 6.3 but for case P-SV-1. There is now one barrier on the fault that remains unbroken at the completion of the dynamic fracture of the fault. (*From Das and Aki*, 1977b. © *Am. Geophys. Union.*)

limit can be obtained by comparing the numerical solution for some simple case with an analytic solution, the spectra in all the cases plotted in this example being shown only up to the frequency where the numerical results are valid) have more energy than that for the smooth fault.

7. The corner frequency averaged over all directions from the source is unaffected by the presence of unbreakable barriers.

8. The stress drop averaged over the total fault length (including the barriers) is lower for the case with unbroken barriers than the other cases. In fact, there is a stress increase on these unbroken regions due to the earthquake. Thus, a complex earthquake with lower average stress drop can generate waves of

6.6   Same as Figure 6.4 but for the case P-SV-1. The dashed lines on the spectra are the curves for the case P-SV-0 and are included for the purpose of comparison. (*From Das and Aki*, 1977b. © *Am. Geophys. Union.*)

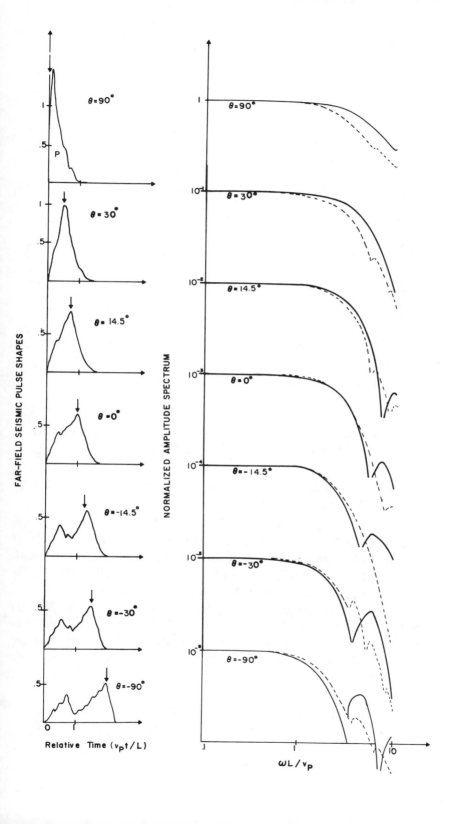

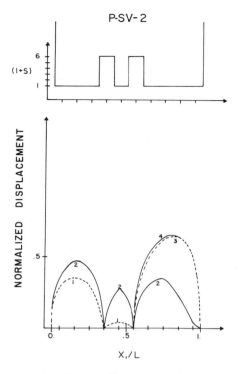

6.7 Same as Figure 6.3 but for case P-SV-2. The two barriers on the fault remain unbroken. (*From Das and Aki*, 1977*b*. © *Am. Geophys. Union.*)

relatively higher frequency than a simple earthquake with relatively higher stress drop.

To consider the effect of fault heterogeneity in estimating corner frequency, let us follow Madariaga (1981) and consider a series of complex pulses, each being the sum of two triangular pulses (a schematization of the pulses obtained above for the barrier model), having a base width of t', say, and separated by some time interval $\Delta t'$, say. For $\Delta t' = 0$, the two pulses coincide, and the triangular pulse and its spectrum are shown at the top of Figure 6.11, the amplitude of the pulse being in arbitrary units. As $\Delta t'$ is increased, the pulses separate; the resulting pulses and their spectra are shown in Figure 6.11 for the cases $\Delta t' = t'/3$, $\Delta t' = 2t'/3$, and $\Delta t' = t'$. The spectra show that the corner

6.8 Same as Figure 6.4 but for the case P-SV-2. The dashed lines give the curve for the case P-SV-0. (*From Das and Aki*, 1977*b*. © *Am. Geophys. Union.*)

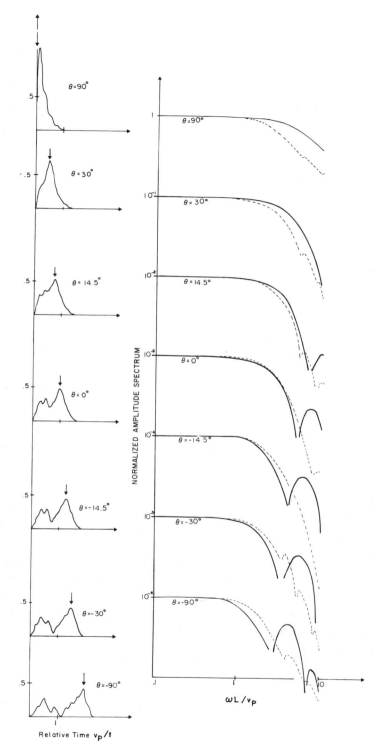

FAR-FIELD SEISMIC PULSE SHAPES

$\theta = 90°$

$\theta = 30°$

$\theta = 14.5°$

$\theta = 0°$

$\theta = -14.5°$

$\theta = -30°$

$\theta = -90°$

Relative Time v_P/t

NORMALIZED AMPLITUDE SPECTRUM

$\theta = 90°$

$\theta = 30°$

$\theta = 14.5°$

$\theta = 0°$

$\theta = -14.5°$

$\theta = -30°$

$\theta = -90°$

$\omega L/v_P$

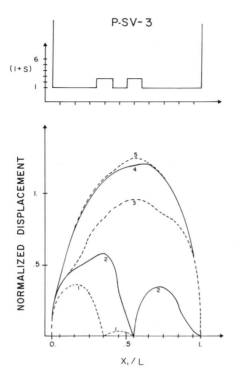

6.9   Same as Figure 6.3 but for case P-SV-3. The two barriers on the fault are of intermediate strength and eventually break while dynamic fracturing of other parts of the fault is continuing. (*From Das and Aki*, 1977b. © *Am. Geophys. Union.*)

frequency decreases slightly for $\Delta t' = t'/3$ from that for $\Delta t' = 0$ but remains unchanged for the other cases. This implies that the corner frequency reflects not the total duration of the time pulse but rather the width of each of the triangles. In other words, the spectra are dominated by the simplest common pulse. The multiplicity of the pulses shows up in the spectral holes that move to lower and lower frequencies as $\Delta t'$ increases. Thus, the corner frequency is not proportional to the overall pulse width for complex earthquakes.

The observational support for complex faulting models came from both seismology and geology. Observations of multiple shocks on seismograms were mentioned at the beginning of this section. The measured

6.10   Same as Figure 6.4 but for case P-SV-3. (*From Das and Aki*, 1977b. © *Am. Geophys. Union.*)

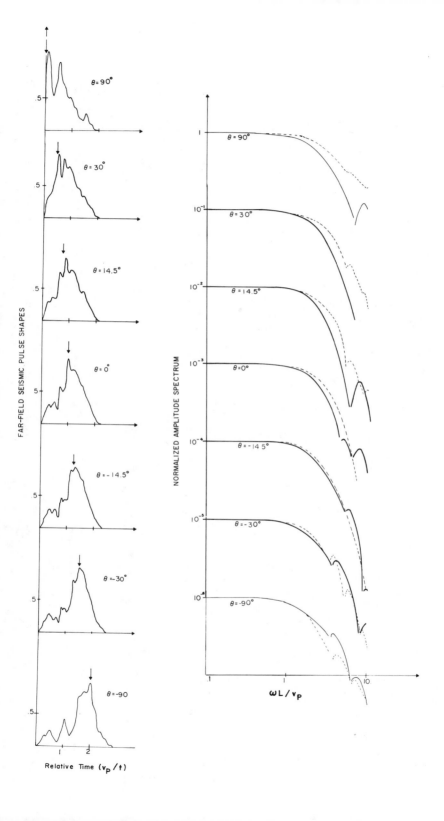

FAR-FIELD SEISMIC PULSE SHAPES

$\theta = 90°$

$\theta = 30°$

$\theta = 14.5°$

$\theta = 0°$

$\theta = -14.5°$

$\theta = -30°$

$\theta = -90$

Relative Time (v_p/t)

NORMALIZED AMPLITUDE SPECTRUM

$\theta = 90°$

$\theta = 30°$

$\theta = 14.5°$

$\theta = 0°$

$\theta = -14.5°$

$\theta = -30°$

$\theta = -90°$

$\omega L / v_p$

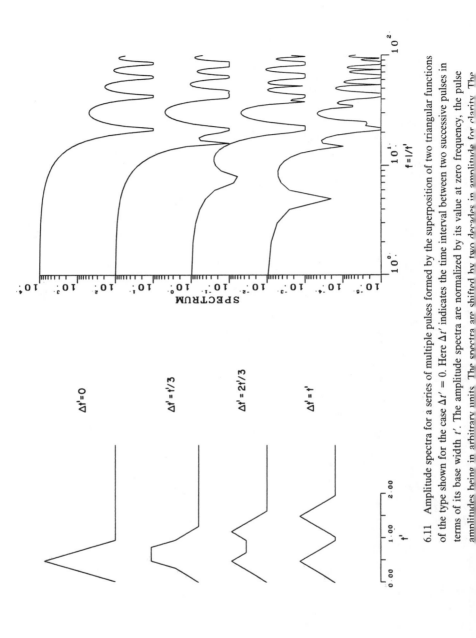

6.11   Amplitude spectra for a series of multiple pulses formed by the superposition of two triangular functions of the type shown for the case $\Delta t' = 0$. Here $\Delta t'$ indicates the time interval between two successive pulses in terms of its base width t'. The amplitude spectra are normalized by its value at zero frequency, the pulse amplitudes being in arbitrary units. The spectra are shifted by two decades in amplitude for clarity. The surface slip after large earthquakes often shows a form similar to the fault slip found for P-SV-1 and P-SV-2. Direct evidence from fractures on mine faces showed that faults are usually very complex, with side steps and highly deformed but unbroken ligaments in the stepover regions (Spottiswoode and McGarr, 1975; McGarr et al., 1979). The impact of this model, in spite of its idealizations, on the understanding of the earthquake faulting process was significant. It led to the characterization of barriers as being material (large S) or geometric (when the fault plane deviated from planarity) by Aki (1979). It also led to the identification of barriers in the field by structural geologists and by seismologists in various locations around the world (Lindh and Boore, 1981; King and Yielding, 1984; Nabelek and King, 1985; Sibson, 1986; Barka and Kadinsky-Cade, in press; Bruhn, Gibler, and Parry, 1987, to name only a few). Major projects are under way in many countries to identify barriers along faults and to try to understand the origin and geochemical characteristics of barriers. The primary reason for this general interest is that earthquakes often nucleate and terminate at barriers.

Since the unbroken barrier with its high residual stress concentration can become the "asperity" of a future earthquake on the same fault, it is important to consider the radiation due to the fracturing of such an unbroken barrier. In Section 5.6, we studied the dynamic fracture of isolated asperities of different shapes on infinite faults. In the next subsection, we will look at the far-field radiation generated by such a model.

Radiation due to the failure of an isolated asperity

The far-field displacement pulse shapes can be conveniently calculated for this case using (6.1.5). The corresponding radiation patterns were given in Section 4.6. Let us consider the far-field pulse shape for the circular asperity along the direction of the normal to the fault. In this direction, the P- and the S-wave pulses coincide. The pulses are given by the term

$$\sum_{i,\,j\in\Sigma_k}\tau_{\alpha i j k'}$$

of (6.1.5) and plotted in Figure 6.12. The normalized time in the figure is $v_\mathrm{p}t/\Delta X$, and it is measured from the time of arrival of the first wave at the receiver. The most striking feature of this pulse is that there is a permanent offset, in contrast to what we saw in the previous examples in

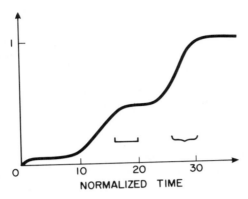

NORMALIZED TIME

6.12   Far-field displacement pulse shape for the P and S waves along the normal to the asperity, due to the fracturing of a circular asperity (Figure 5.18). The bracket and brace are explained in the text. (*From Das and Kostrov, 1983. © Am. Geophys. Union.*)

this chapter for the conventional crack model. This is not surprising when we recall that the problem was formulated such that the two half-spaces on either side of the fault plane remain permanently shifted after the asperity has fractured and disappeared. The rise time of the displacement from zero to this final value is the time required for the asperity to fracture. The brace in the figure indicates the time when the number of grids broken per unit time is the highest. The fracture process for this case (Figure 5.18) shows that this indeed took place toward the end of the breaking process. The square bracket indicates the period when the breaking rate is high but the displacement does not increase. This is because, although the number of grids breaking per unit time is large, these points are situated far from one another on the asperity; also, they do not have large stress drops associated with them and hence do not contribute significantly to the increase in the far-field displacement. If we looked at the acceleration pulse shape (obtained simply by twice differentiating the displacement pulse in Figure 6.12), we would find that the high accelerations correspond in time to the (relative) times when the breaking rate of grids on the asperity is the highest. Thus, the far-field displacements are very sensitive to the location of fracturing points on the asperity, whereas the accelerations are sensitive to the rate of increase of the broken area but not to its distribution over the fault. The pulse shapes in other directions from the source have essentially similar characteristics, the rise times being shortest in the (general) direction of fracture propagation and longest in the opposite direction.

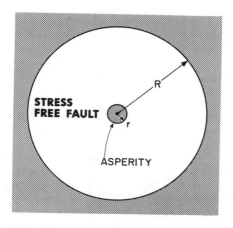

6.13   Geometry of the problem of failure of an isolated asperity on a finite fault.

The pulses for elliptical asperity fracture (Das and Kostrov, 1985) are similar and are not included here.

The asperity model

The basic idea of this model was suggested by Madariaga (1979) and by Rudnicki and Kanamori (1981). According to the model, an earthquake is caused by the failure of isolated, highly stressed regions of the fault, the rest of the fault having little or no resistance to slip (being partially broken and preslipped, say) and contributing little or no stress drop to the earthquake process. This results in a nonuniform stress drop over the fault. Since the regions without slip are able to withstand the high stresses concentrated on it until the moment the earthquake begins, it must be assumed that the parameter σ^u for these regions is higher than that for the rest of the fault. The spontaneous, dynamic fracturing of one or more such isolated asperities of general shape and size on a finite fault has not yet been studied. The simpler problem of radiation from the fracturing of a circular asperity at the center of a circular fault was studied by Das and Kostrov (1986), and we shall discuss the result here.

In this model, a circular crack of radius R, say, has a circular asperity of radius r, say, at its center (Figure 6.13). The annular region between the crack and the asperity is broken and assumed to be at or very close to the kinetic frictional level. When the central asperity breaks, this annular region exhibits no (or little) dynamic stress drop. It also has little or no resistance to slip. For the numerical calculations, r/R is taken as

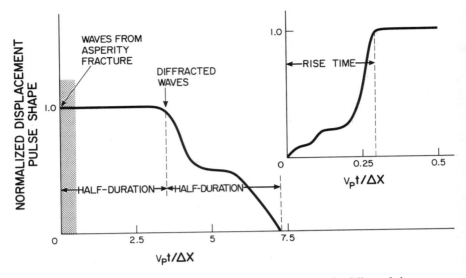

6.14   Far-field S-wave displacement pulse shape due to the failure of the
isolated asperity shown in Figure 6.12. Inset shows details of pulse shape in
stippled region. (*From Das and Kostrov, 1986. © Am. Geophys. Union.*)

.1 and the asperity is taken as a single spatial grid. The asperity is
released, and the ensuing dynamic slip is allowed to spread out over the
entire circular fault. The slip is calculated using algorithm (5.2.13) and
the P- and S-wave pulse shapes in different directions from the source
are found using (6.1.4). The normalized S-wave displacement pulse shape
looking down at the fault along the normal as a function of normalized
time $v_p t/\Delta X$ is shown in Figure 6.14 as a representative example. The
displacement pulse immediately reaches its maximum value and remains
flat until the first diffracted waves from the crack edge arrive at the
observer at time $T_d \simeq R/v_S$, measured from the time of arrival of the
first S wave. The displacement then starts decreasing and finally reaches
zero at time $\simeq 2T_d$ ($\simeq 2R/v_R$). The minor oscillations following this
that occur due to backslip being permitted on the fault are ignored in
this figure. The duration of the flat part of the pulse is thus controlled by
the size of the large crack of radius R. Since the asperity was released
instantaneously in this problem, this picture does not represent the rising
part of the pulse correctly. But this was calculated in the last subsection,
and using those results and adjusting the time scale, we obtain the pulse
shape in the stippled region of Figure 6.14, as shown in the inset of this
figure. The rise time for the failure of a single asperity was shown in the

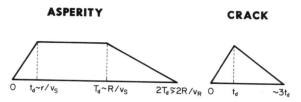

6.15   Schematic representation of the far-field displacement pulse due to the failure of an asperity on a finite fault and due to a propagating crack. (*From Das and Kostrov, 1986. © Am. Geophys. Union.*)

last subsection to be controlled by the asperity size and is given by $t_d \approx r/\beta$.

Since the details of the pulses of Figure 6.14 depend on the parameters of the particular problem (crack and asperity size and shape, fracture velocity, etc), one may neglect the details and construct a schematic representation of the pulse shape due to the fracture of an isolated asperity on a finite fault, as shown in Figure 6.15. The main features of this pulse are a steeply rising part followed by a flat portion of long duration and then a gradual return of the displacement pulse to zero. The triangular pulse from a circular crack (Figure 6.1) is also shown in the figure for comparison. If this circular crack is taken to be the same size as the asperity of the model under discussion here, then the pulse shape would have the same rise time as the asperity model. However, once the maximum amplitude is reached, the two pulses become very different in character, the crack pulse immediately starting to decrease toward zero and reaching zero at time about twice the rise time, as we saw earlier in this chapter.

Thus, the asperity pulse has an anomalously large seismic moment (area under the pulse) and anomalously large duration compared with a crack pulse for a crack of radius r. Such earthquakes have been called "slow" or "weak" earthquakes (Kanamori and Cipar, 1974; Kuznetsova et al., 1976). The spectrum of the pulse shown in Figure 6.14 was found to have the same general form as that for the conventional crack model (Figure 6.1).

It must be pointed out here that this model is not the only possible model for slow earthquakes. Clearly, such earthquakes could also be modeled as a very slowly propagating crack, due, for example, to very low stress drop.

Next let us compare the seismic moments [defined in equation (4.1.20)] for the crack and asperity models. The seismic moment M_0^{cr} for a circular crack with average stress drop $\Delta\sigma$ is given by

$$M_0^{cr} = \tfrac{16}{7}\Delta\sigma R^3 \qquad (6.3.1)$$

for a Poisson solid. To determine the seismic moment for the asperity problem, let us assume that the earlier stress drop on the now fractured annular region of the fault is also $\Delta\sigma$. This stress is now concentrated on the unbroken central asperity. If the asperity were absent, the slip at the site of the asperity would be given by the static crack solution

$$[u_0] = \frac{24}{7\pi}\frac{\Delta\sigma}{\mu}R \qquad (6.3.2)$$

Assuming that the asperity radius $r \ll R$, the stress distribution $\Delta\sigma(\rho)$ on the asperity will be approximately the same as that for an asperity on an infinite fault loaded by the same amount of remote slip so that

$$\Delta\sigma(\rho) = \frac{8}{7\pi}\frac{\mu[u_0]}{\sqrt{r^2 - \rho^2}} \qquad (6.3.3)$$

where ρ is the distance from the center of the asperity. The average stress drop $\overline{\Delta\sigma}$ on the asperity can be found by integrating equation (6.3.3) over the asperity:

$$\frac{\overline{\Delta\sigma}}{\Delta\sigma} = .8\frac{R}{r} \approx \frac{R}{r} \qquad (6.3.4)$$

The average stress drop on the asperity is thus increased by the ratio R/r over the average stress drop that had previously occurred over the annular region of the fault. Applying Betti's theorem to the asperity problem under consideration and to the problem of a circular crack of radius R with uniform stress drop $\Delta\sigma$, we obtain

$$\Delta\sigma \int_{S^R} [u^{\text{asp}}]\, dS = [u_0]\int_{S^R} \Delta\sigma(\rho)\, dS \qquad (6.3.5)$$

where $[u^{\text{asp}}]$ denotes the slip due to asperity failure and S^R denotes the total fault area. Substituting (6.3.2) and (6.3.3) into (6.3.5), we get

$$M_0^{\text{asp}} = 1.19\frac{16}{7}\Delta\sigma R^3\frac{r}{R} \qquad (6.3.6)$$

where M_0^{asp} is the seismic moment of the asperity problem considered in this section. Replacing the constant 1.19 by unity for the purposes of estimation, we get

$$M_0^{\text{asp}} \approx M_0^{cr}\left(\frac{r}{R}\right), \qquad \text{i.e.,} \qquad M_0^{\text{asp}} \approx rR^2\,\Delta\sigma \approx Rr^2\,\overline{\Delta\sigma} \qquad (6.3.7)$$

Thus, the seismic moment due to fracture of a single asperity of radius r on a finite, stress drop free fault of radius R is r/R times smaller than the seismic moment due to a circular fault of radius R with uniform stress drop $\Delta\sigma$ but $(R/r)^2$ larger than the seismic moment due to a fault of radius r with stress drop $\Delta\sigma$.

To recognize such "slow" or "weak" earthquakes in practice, let us compare the seismic moments and corner frequencies for the crack and asperity model for earthquakes of the *same magnitude*. The magnitude M, say, is proportional to the logarithm of the displacement spectral density at a fixed frequency ω_M (where $\omega_M/2\pi = 1$ Hz for body wave magnitudes and corresponds to 20-sec period waves for surface wave magnitudes). For the sake of simplicity, let us confine our discussion to the case in which ω_M is greater than any corner frequency in the problem. Using the conventional parameterization of the displacement spectral density envelope, we have

$$M = \log M_0 - 2\log \omega_M + 2\log \omega_c + C \tag{6.3.8}$$

where C is some constant and ω_c is the corner frequency. Taking logarithm of (6.3.1) and/or (6.3.7), we get

$$\log M_0^{\text{asp}} = -3\log \omega_c + \log \alpha + C_1 \tag{6.3.9}$$

where $\alpha = 1$ for crack models, $\alpha = r/R$ for the asperity model, and ω_c is taken proportional to $1/R$ in both cases. The constants C and C_1 are essentially the same for both models. Using equations (6.3.8) and (6.3.9) for the two models at a given magnitude M, we obtain

$$\log M_0^{\text{asp}} = \log M_0^{\text{cr}} - 2\log(r/R) \tag{6.3.10}$$

and

$$\log \omega_c^{\text{asp}} = \log \omega_c^{\text{cr}} + \log(r/R) \tag{6.3.11}$$

or

$$M_0^{\text{asp}} = M_0^{\text{cr}}\left(\frac{R}{r}\right)^2 \quad \text{and} \quad \omega_c^{\text{asp}} = \omega_c^{\text{cr}}\left(\frac{r}{R}\right) \tag{6.3.12}$$

Thus, an earthquake due to asperity failure has an $(R/r)^2$ times larger seismic moment and (r/R) times lower corner frequency than the earthquake due to a crack of the *same magnitude*. This is illustrated in Figure 6.16. A similar comparison can easily be made for a crack and an asperity earthquake having the same seismic moment.

The relations between moment and magnitude for the crack and the asperity models provide a comparison between the long-period (moment) and the short-period (magnitude) radiation characteristics of the two

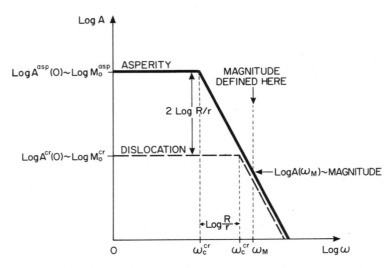

6.16   Far-field displacement spectra for the isolated asperity failure and the propagating crack models, showing the relation between seismic moment and corner frequency for the two models *at a given magnitude*.

source models. If one has reliable plots of corner frequency and/or seismic moment against magnitude for "normal" (crack) earthquakes, then for a given magnitude one can read off the expected values of ω_c^{cr} and/or M_0^{cr} from the plots. If, then, the observed values of corner frequency and/or seismic moment of the next earthquake fall off these curves beyond a few standard deviations, one may conclude that the earthquake is a slow one. If it is assumed that the slow earthquake is due to an asperity model of the kind discussed here and ω_c^{asp} and M_0^{asp} are its observed corner frequency and seismic moment, then the ratio of the dimension of the asperity r to the whole fault R would be given by

$$\left(\frac{r}{R}\right)^2 = \frac{M_0^{cr}}{M_0^{asp}} \quad \text{or} \quad \frac{r}{R} = \frac{\omega_c^{asp}}{\omega_c^{cr}} \qquad (6.3.13)$$

As an example of these results, let us consider that the outer crack corresponds to an active fault area sufficiently damaged during the preseismic period, and let a foreshock be due to the fracture of one of the few remaining asperities. Then the moment of such a foreshock would be amplified by the square of the ratio of the impending earthquake fault size to the size of this asperity. Furthermore, the corner frequency of such a foreshock would be lowered according to equation

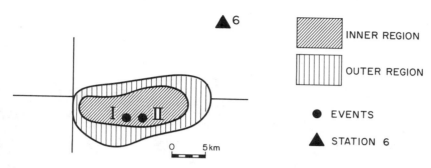

6.17 Fault geometry and station location for two earthquakes near Garm, USSR. The faulted area, the foci of the earthquake, and the "inner" and "outer" areas of the fault where many smaller earthquakes occur are indicated. Earthquake I occurred on September 3, 1976 ($m_b \approx 5.0$) and earthquake II occurred on December 25, 1977 ($m_b \approx 5.0$). The station marked 6 is the station where the corner frequency observations plotted in Figure 6.18 were made. The smaller events had magnitudes $m_b \approx 2.0$–3.0. (*Modified from Martynov, 1983.*)

(6.3.12). Such a situation has been reported in the literature by Martynov (1983), who studied the corner frequencies of hundreds of foreshocks in Garm, USSR, from 1972 to 1978, which preceded two larger events of $m_b \approx 5.0$. The geometry of the fault and the station location are shown in Figure 6.17. The fault is divided into inner and outer regions, the inner region surrounding the main shock foci, marked I and II. The S-wave corner frequencies are plotted against time for events in the inner and outer areas of the fault in Figure 6.18. Each data point (solid and open circles correspond to events in the inner and outer regions, respectively) represents the average of several events, the standard deviation in corner frequency being indicated by bars. A significant decrease in corner frequency is seen for events in the inner region before Event I and a less clear decrease is seen before Event II. This decrease was much larger than the standard deviations and persisted for more than a year before Event I and for a short time following it. On the basis of the asperity model discussed above, the following explanation may be suggested for this observation. The earlier foreshocks before Event I occur on an essentially unfractured fault and are well modeled as cracks. As the time of occurrence of I approaches, preslip takes place on the regions of the fault close to the hypocenter (inner zone), so that the foreshocks occurring in this region represent earthquakes due to asperity failure. The events in the outer region are close to the edge of the main fault and

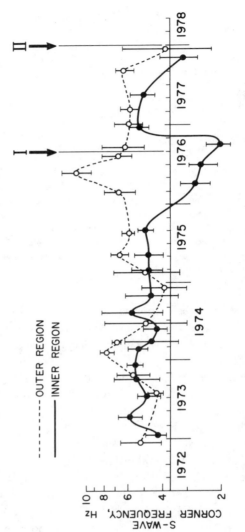

6.18 Observed S-wave corner frequency versus time for events occurring within the faulted areas of earthquakes I and II. (*Modified from Martynov, 1983.*)

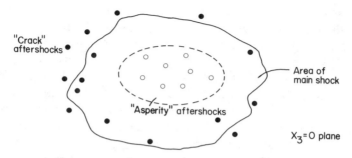

6.19   Schematic diagram showing the difference in mechanical conditions on a fault for aftershocks located in the interior and for those situated near its edges.

still appear as normal events. The shocks following I can be considered aftershocks of I or foreshocks of II but in either case are events caused by the failure of locked regions surrounded by unlocked regions of the fault.

Radiation characteristics of aftershocks

The most convincing evidence that faults are heterogeneous not only near the surface of the earth but also at the depths where the main faulting in an earthquake occurs is that aftershocks do occur at these depths. The asperity model of the last subsection not only is a model for large earthquakes but is also clearly a model of aftershocks that occur within the fractured area of the main shock. Thus, two types of aftershock occur on the main fault plane (aftershocks also occur off the fault, as we mentioned earlier in this section). The aftershocks that occur within the main shock area are regions that may have remained unbroken during the main shock and fracture later due to increased loading on them combined with some time-dependent effects (e.g., static fatigue). Such aftershocks must appear as "low-frequency, low-stress-drop" events. The aftershocks that occur beyond the main fault edge are due to the breaking of regions that are loaded by the increased concentration of stress at the fault edge, and their fracture areas are surrounded by unslipped regions of the fault plane. Thus, the interior and the edge aftershocks are mechanically different and require different initial and boundary conditions on the fault. The exterior aftershocks can be modeled by the conventional crack model. The two types of aftershock geometry are shown in Figure 6.19.

From the results of the last subsection, the seismic moment of an interior aftershock due to the failure of an asperity of radius r is less than the moment of the main shock of radius R by the factor r/R but larger than the moment of an exterior aftershock of radius r by the factor $(R/r)^2$. An example of this is given in Kuznetsova et al. (1976). These authors studied the aftershocks of the earthquake that took place on May 14, 1970, in Dagestan, USSR ($M_s = 6.6$). They observed that the aftershocks near the central part of the faulted zone had an anomalously low ratio of seismic energy to seismic moment, whereas the aftershocks near the peripheral part had normal values of this ratio.

In the light of this observation, the idea of using aftershocks to obtain the medium response for synthesizing the main shock (Hartzell, 1978; Joyner and Boore, 1986) must be somewhat modified to account for the difference in spectral content of aftershocks located on different parts of the fault plane.

APPENDIXES

App. 1 Kernels of the integral representations (5.1.9) and (5.1.3)

The kernel $G_{\alpha\beta}(\mathbf{X}, t)$ of the representation relation (5.1.9) can be written in closed form for the three-dimensional problem in the upper half-space $X_3 \geq 0$, as follows:

$$G_{11}(\mathbf{X}, t) = -\frac{1}{\pi \mu r} \frac{d}{dt} \left[I_1 \left(\frac{v_P t}{r} \right) \cos^2 \phi - I_2 \left(\frac{v_P t}{r} \right) \sin^2 \phi \right]$$

$$G_{12}(\mathbf{X}, t) = G_{21}(\mathbf{X}, t)$$

$$= -\frac{1}{\pi \mu r} \frac{d}{dt} \left[I_1 \left(\frac{v_P t}{r} \right) + I_2 \left(\frac{v_P t}{r} \right) \right] \sin \phi \cos \phi \qquad (A1.1)$$

$$G_{22}(\mathbf{X}, t) = -\frac{1}{\pi \mu r} \frac{d}{dt} \left[I_1 \left(\frac{v_P t}{r} \right) \sin^2 \phi - I_2 \left(\frac{v_P t}{r} \right) \cos^2 \phi \right]$$

where the two-dimensional vector $\mathbf{X} \equiv (r, \phi)$, its polar coordinates. For a Poisson solid, I_1 and I_2 are given by

$$I_1(T) = \begin{cases} 0 & \text{for} \quad T < 1 \\ \begin{aligned} T^2 \big\{ & c_1/(T^2 - R_1)^{1/2} \\ & -c_2/(T^2 - R_2)^{1/2} \\ & -c_3/(R_3 - T^2)^{1/2} \big\} \end{aligned} & \text{for} \quad 1 < T < (v_P/v_S) \\ 0.5 - 2T^2 c_3/(R_3 - T^2)^{1/2} & \text{for} \quad (v_P/v_S) < T < \sqrt{R_3} \\ & \qquad\qquad = (v_P/v_R) \\ 0.5 & \text{for} \quad \sqrt{R_3} < T \end{cases}$$

259

$$I_2(T) = \begin{cases} 0 & \text{for } T < 1 \\ \begin{aligned} & -c_4 + c_1(T^2 - R_1)^{1/2} \\ & \quad -c_2(T^2 - R_2)^{1/2} \\ & \quad +c_3(R_3 - T^2)^{1/2} \end{aligned} & \text{for } 1 < T < (v_P/v_S) \\ -2c_4 + 2c_3(R_3 - T^2)^{1/2} & \text{for } (v_P/v_S) < T < \sqrt{R_3} \\ -2c_4 & \text{for } \sqrt{R_3} < T \end{cases}$$

and v_S is the shear wave speed of the medium. The constants R_1, R_2, and R_3 are solutions to the Rayleigh cubic equation in ξ^2 given by (3.5.13) and, in particular, $R_3 = (v_P/v_R)^2$, where v_R is the Rayleigh wave speed. The positive constants c_1, c_2, c_3, and c_4 are given by

$$c_1 = -2a\frac{v_P^2}{v_S^2}\left(\frac{v_P^2}{v_S^2} - R_1\right)(1 - R_1)^{1/2}$$

$$c_2 = 2b\frac{v_P^2}{v_S^2}\left(\frac{v_P^2}{v_S^2} - R_2\right)(1 - R_2)^{1/2}$$

$$c_3 = -2c\left(\frac{v_P^2}{v_S^2}\right)\left(R_3 - \frac{v_P^2}{v_S^2}\right)(R_3 - 1)^{1/2}$$

$$c_4 = \frac{v_P^2/v_S^2}{(8v_P^2/v_S^2) - 8}$$

where a, b, and c are given by

$$a^{-1} = 16\left(\frac{v_P^2}{v_S^2} - 1\right)(R_1 - R_2)(R_3 - R_1)$$

$$b^{-1} = 16\left(\frac{v_P^2}{v_S^2} - 1\right)(R_1 - R_2)(R_2 - R_3)$$

$$c^{-1} = 16\left(\frac{v_P^2}{v_S^2} - 1\right)(R_3 - R_1)(R_2 - R_3)$$

In Figure A1.1 are plotted I_1, I_2, dI_1/dt, and dI_2/dt. These Green functions were obtained by Chao (1960). Richards (1979) obtained the solution for a solid with general Lamé parameters λ, μ.

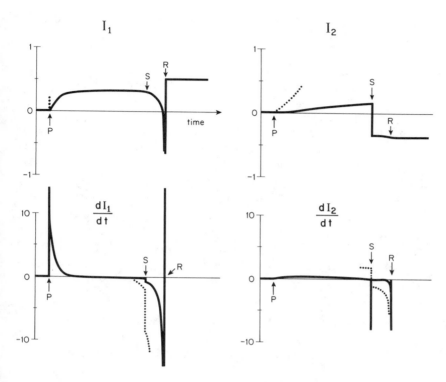

A1.1   Plots of the functions I_1, I_2 (*modified from Richards, 1979*) and dI_1/dt, dI_2/dt against ($v_P t/r$), for a Poisson solid. The dotted lines are the values of the functions increased tenfold to show detailed behavior at low amplitudes. The letters P, S, and R denote the arrival times of the P, S, and Rayleigh waves, respectively. Values are plotted as heavy lines only between amplitudes ± 1 for I_1 and I_2 and between amplitudes ± 10 for dI_1/dt and dI_2/dt. Function I_1 is singular at R, dI_1/dt is singular at P and R, and dI_2/dt is singular at S and R. (*After Das*, 1980. © *Roy. Astr. Soc.*)

The tensor $G_{\alpha\beta}$ is a homogeneous function in X_1, X_2, and t of the order -2 and vanishes for $t < 0$. More precisely, $G_{\alpha\beta}$ vanishes outside the characteristic cone, that is, in the region $v_P^2 t^2 > X_1^2 + X_2^2$ (by the principle of causality). We shall call this cone the "P cone." Furthermore, $G_{\alpha\beta}$ vanishes once the Rayleigh wave has passed, that is, in the region $v_R^2 t^2 < X_1^2 + X_2^2$. We shall call this cone the "Rayleigh cone." Expressions (A1.1) show that $G_{\alpha\beta}$ has the following properties: G_{11} and G_{22} are 90° rotations of one another and are symmetric about the X_1 and X_2 axes. G_{12} has an eightfold symmetry, being symmetric about $\phi = 45°$, 135°, and so on and is antisymmetric about the X_1 and X_2

axes. The integral of $G_{\alpha\beta}$ over the entire $X_3 = 0$ plane at any fixed time t is independent of t, that is,

$$\iint_{-\infty}^{\infty} G_{\alpha\beta}(X_1, X_2, t)\, dX_1\, dX_2 = \text{const} \qquad \text{if } \alpha = \beta$$
$$= 0 \qquad \text{if } \alpha \neq \beta \tag{A1.2}$$

The discrete Green function $F_{\alpha\beta}$ needed in (5.2.9) is obtained using (5.2.4). At least one of the three integrations involved in (5.2.4) must be performed numerically, but since $G_{\alpha\beta}$ has only integrable singularities, this is simple. $F_{\alpha\beta}$ conserves many important properties of $G_{\alpha\beta}$; namely, $F_{\alpha\beta}$ is nonvanishing only between the discrete counterparts of the P and Rayleigh cones, has the same symmetries as $G_{\alpha\beta}$ relative to the coordinate axes, and satisfies a relation similar to (A1.2), with the integrals replaced by summations. For $v_P(\Delta t - \delta t) \leq \Delta X/2$, $F_{\alpha\beta}(i, j, 0)$ will be nonvanishing only for $i = 0 = j$. The value of this tip element of the P cone is then

$$F_{\alpha\beta}(0,0,0) = -2\int_0^{\delta t} dt \iint_{-\infty}^{\infty} G_{\alpha\beta}(\xi_1, \xi_2, \tau)\, d\xi_1\, d\xi_2$$

Using (A1.2) and the homogeneity and symmetry properties of $G_{\alpha\beta}$,

$$F_{\alpha\beta}(0,0,0) = \delta t F_0 \delta_{\alpha\beta}$$

where

$$F_0 = -2\int_0^1 d\tau \iint_{-\infty}^{\infty} G_{11}(\xi_1, \xi_2, \tau)\, d\xi_1\, d\xi_2$$

is a positive constant independent of the grid size and $\delta_{\alpha\beta}$ is the Kronecker delta. $F_{\alpha\beta}(0,0,0)$ is the maximum element of $F_{\alpha\beta}$. The minus signs were introduced to make F_0 positive. (The sign of F_0 becomes important in the problems of cracks with friction discussed in Section 5.5.) Furthermore, from (A1.2), the sum of $F_{\alpha\beta}(i, j, k)$ for a fixed $k > 0$ over all values of i and j is equal to $\Delta t F_0 \delta_{\alpha\beta}$, that is,

$$\sum_{i=-\infty}^{\infty} \sum_{j=-\infty}^{\infty} F_{\alpha\beta}(i, j, k) = \Delta t F_0 \delta_{\alpha\beta} \tag{A1.3}$$

$F_{\alpha\beta}$ calculated by Das (1980) and by Das and Kostrov (1987) automatically satisfies this condition to several decimal places. In performing the numerical integration to obtain $F_{\alpha\beta}$ from $G_{\alpha\beta}$ special attention must be paid to the times and places near the arrivals of the P, S, and Rayleigh

$$F_{\alpha\alpha}(X_1, X_2, 40)$$

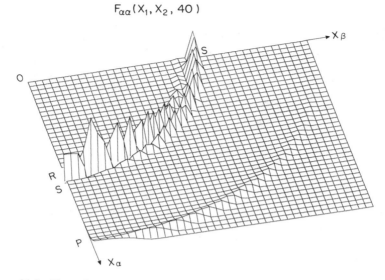

A1.2 Three-dimensional plot of the discretized Green function component $F_{\alpha\alpha}(X_1, X_2, 40)$ at normalized time $v_p t / \Delta X = 40$. X_α is the direction of the impulse applied at 0, and X_β is the orthogonal axis on the $X_3 = 0$ plane. The letters P, S, and R denote the P, S, and Rayleigh waves, respectively.

wave fronts. Figure A1.2 shows the $F_{\alpha\alpha}$ component of the discrete Green function in a perspective plot.

The kernel $T_{\alpha\beta}$ of the representation relation (5.1.3) has nonintegrable singularities, and its discrete counterpart $S_{\alpha\beta}$ was obtained in Chapter 5 by inverting $F_{\alpha\beta}$. The general properties of the discrete kernel $S_{\alpha\beta}$ are similar to those of $F_{\alpha\beta}$. That is, $S_{\alpha\beta}$ has the same symmetries as $F_{\alpha\beta}$ and satisfies a relation of the form (A1.3) with F_0 replaced by the corresponding tip value $S_0 = S_{11}(0,0,0) = 1/\Delta t G_0$, where G_0 is a positive constant satisfying (5.2.10). G_0 is discussed further in Appendix 2. S_0 is the maximum element of $S_{\alpha\beta}$. The elements of $F_{\alpha\beta}$ are zero for grids situated totally within the Rayleigh cone. The analytic kernel $T_{\alpha\beta}$ vanishes within the even wider "S cone" given by $v_s^2 t^2 = X_1^2 + X_2^2$. This property is not conserved in its discrete counterpart $S_{\alpha\beta}$. For a given k, the maximum values of it are observed at or near the axis $i = 0 = j$. In general, $S_{\alpha\beta}$ does not approximate the dislocation kernel $T_{\alpha\beta}$. In fact, the integral operator in (5.1.3) is a degenerate one and can be inverted only under additional conditions, as mentioned in Section 5.1, whereas $S_{\alpha\beta}$ is a well-defined matrix and can be inverted without any reservations.

For completeness, we reproduce the Green function $G_{\alpha\beta}$ for the two-dimensional shear problem. For the antiplane problem, let the crack extension be in the X_1 direction, the only nonzero initially applied stress component being σ_2^0. The slip component $a_2(X_1, t)$ is given, from (5.1.9), as

$$a_2(X_1, t) = 2\int_0^t dt' \int_L G_{22}(X_1 - X', t - t')\tau_2(X', t')\, dX'$$

where the Green function G_{22} is given by

$$G_{22}(X, t) = \frac{-v_S H[v_S t - X]}{\pi\mu R} \tag{A1.4}$$

$H[\ ]$ being the Heaviside function and $R = \sqrt{v_S^2 t^2 - X^2}$. The region of integration L is the triangular portion of the $(X_1 - t)$ plane given by

$$v_S^2(t - t') - (X_1 - X')^2 \geq 0, \qquad t \geq t' \geq 0$$

For the inplane shear problem, let the crack extend in the X_1 direction, the direction of the initial stress being σ_1^0. The slip component $a_1(X_1, t)$ is obtained from (5.1.9) as

$$a_1(X_1, t) = 2\int_0^t dt' \int_L G_{11}(X_1 - X', t - t')\tau_1(X', t')\, dX'$$

where L is now the triangular portion of the $(X_1 - t)$ plane given by

$$v_P^2(t - t') - (X_1 - X')^2 \geq 0, \qquad t \geq t' \geq 0$$

The Green function G_{11} was given by Lamb (1904) as

$$G_{11}(X, t) = -\frac{1}{\pi\mu X v_S^2}$$

$$\times \left\{ 4T^2\left(T^2 - \frac{1}{v_S^2}\right)\sqrt{T^2 - \frac{1}{v_P^2}}\, H\left[T - \frac{1}{v_P}\right] \right.$$

$$\left. + \left(2T^2 - \frac{1}{v_S^2}\right)\sqrt{T^2 - \frac{1}{v_S^2}}\, H\left[T - \frac{1}{v_S}\right] \right\} \Big/ h(T^2)$$

$$\tag{A1.5}$$

where now $T = t/X$, and $h(T^2)$ is given by

$$h(T^2) = \left(2T^2 - \frac{1}{v_S^2}\right)^4 - 16T^4\left(T^2 - \frac{1}{v_S^2}\right)\left(T^2 - \frac{1}{v_P^2}\right)$$

For the two-dimensional case, the (two) integrations needed in (5.2.4) to obtain $F_{\alpha\alpha}$ can be performed analytically. We give below the expressions for F_{11} and F_{22} in closed form.

From (5.2.4)

$$F_{11}(i, k) = -2\int_0^{\Delta t} dt \int_{-\Delta X/2}^{\Delta X/2} G_{11}(i\,\Delta X + \xi_1, k\,\Delta t + \delta t - \tau)\, d\xi_1$$

where the integrals over G_{11} can be written as

$$\int_{t_1}^{t_2} dt \int_{X_1}^{X_2} G_{11}(X, t)\, dX$$

$$= \frac{1}{\pi\mu}\left[X_2\varepsilon_1\left(\frac{v_P t_2}{X_2}\right) - X_2\varepsilon_1\left(\frac{v_P t_1}{X_2}\right) - X_1\varepsilon_1\left(\frac{v_P t_2}{X_1}\right) + X_1\varepsilon_1\left(\frac{v_P t_1}{X_1}\right)\right]$$

and $\varepsilon_1(p)$, where $p = v_P t/X$, is given by

$$\varepsilon_1(p) = \frac{v_P^2}{v_S^2}\sum_k a_k\left[b_k\Gamma(q, r_k, 1) + c_k\Gamma\left(q, 0, \frac{v_P^2}{v_S^2}\right) + d_k\Gamma\left(q, r_k, \frac{v_P^2}{v_S^2}\right)\right]$$

where $q = v_P^2 t^2/X^2$, $a_k^{-1} = 16(v_P^2/v_S^2 - 1)(r_n - r_l)(r_m - r_n)$ with l, m, n in cyclic order,

$$b_k = 2\left(r_k - \frac{v_P^2}{v_S^2}\right)(r_k - 1)$$

$$c_k = \frac{(v_P/v_S)^6}{2r_k}$$

$$d_k = \frac{\left((v_P^2/v_S^2) - 2r_k\right)^2\left(r_k - (v_P^2/v_S^2)\right)}{2r_k}$$

and the r_k's are the roots of the equation

$$h(q) = \left(2q - \frac{v_P^2}{v_S^2}\right)^4 - 16\left(\frac{v_P}{v_S}\right)^8 (q - 1)\left(q - \frac{v_P^2}{v_S^2}\right)$$

The Γ's are given by

$$\Gamma(q, r, a) = 0 \qquad\qquad \text{for} \quad q < a$$

$$= \left[2\left(\frac{a}{a-r}\right)^{1/2} \frac{Z\left(1 - \frac{r}{a}\sin^2 Z\right)^{1/2}}{\cos Z} \right.$$

$$-\log \frac{\left(1 - \frac{r}{a}\sin^2 Z\right)^{1/2} + \left(\frac{a-r}{a}\right)^{1/2}\sin Z}{\left(1 - \frac{r}{a}\sin^2 Z\right)^{1/2} - \left(\frac{a-r}{a}\right)^{1/2}\sin Z}$$

$$\left. -2\left(\frac{r}{a-r}\right)^{1/2}\sin^{-1}\left(\left(\frac{r}{a}\right)^{1/2}\sin Z\right) \right]_{Z = \tan^{-1}((1-a)/(a-r))^{1/2}}^{Z = \tan^{-1}((p^2-a)/(a-r))^{1/2}}$$

$$\text{for} \quad q > a, r < a$$

$$= \frac{1}{(r-a)^{1/2}}\left[(Z^2 + a)^{1/2}\log\left(\frac{|Z - (r-a)^{1/2}|}{Z + (r-a)^{1/2}}\right) \right.$$

$$-2(r-a)^{1/2}\log\left(Z + (Z^2 + a)^{1/2}\right)$$

$$\left. + r^{1/2}\log\left|\frac{\frac{(Z^2 + a)^{1/2} + r^{1/2}}{Z - (r-a)^{1/2}} + \left(1 - \frac{a}{r}\right)^{1/2}}{\frac{(Z^2 + a)^{1/2} + r^{1/2}}{Z + (r-a)^{1/2}} - \left(1 - \frac{a}{r}\right)^{1/2}}\right| \right]_{Z = (1-a)^{1/2}}^{Z = (p^2-a)^{1/2}}$$

$$\text{for} \quad q > a, r > a$$

The tip value $F_{11}(0, 0)$ can be obtained simply to be given by

$$F_{11}(0, 0) = v_S/\mu \qquad\qquad\qquad (A1.6)$$

From (5.2.4),

$$F_{22}(i, k) = -2\int_0^{\Delta t} d\tau \int_{-\Delta X/2}^{\Delta X/2} G_{22}(i\,\Delta X + \xi_1, k\,\Delta t + \delta t - \tau)\,d\xi_1$$

where the integrals over G_{22} can be written in closed form as

$$\int_{t_1}^{t_2} dt \int_{X_1}^{X_2} G_{22}(X, t) \, dX = \frac{1}{\pi\mu} \left[X_2 \varepsilon_2 \left(\frac{v_S t_2}{X_2} \right) - X_2 \varepsilon_2 \left(\frac{v_S t_1}{X_2} \right) \right.$$

$$\left. - X_1 \varepsilon_2 \left(\frac{v_S t_2}{X_1} \right) + X_1 \varepsilon_2 \left(\frac{v_S t_1}{X_1} \right) \right]$$

and

$$\varepsilon_2 \left(\frac{v_S t}{X} \right) = \frac{v_S t}{X} \tan^{-1} \sqrt{\frac{v_S^2 t^2}{X^2} - 1}$$

$$- \log \left(\frac{v_S t}{X} + \sqrt{\frac{v_S^2 t^2}{X^2} - 1} \right) \quad \text{for} \quad \frac{v_S^2 t^2}{X^2} > 1$$

$$= 0 \quad \text{for} \quad \frac{v_S^2 t^2}{X^2} < 1$$

The tip value $F_{22}(0, 0)$ is obtained directly from the above formulas to be

$$F_{22}(0, 0) = +v_S/\mu \tag{A1.7}$$

Note that the tip values $F_{\alpha\alpha}(0, 0)$ for the two-dimensional problems are the same. (We shall show in Appendix 2 that it is the same for the three-dimensional problem as well.) Care must be taken when evaluating $\varepsilon_1(p)$ near $p = 1$, and double-precision arithmetic must be used to avoid loss of accuracy in some terms of the expression.

The analytic form of $T_{\alpha\beta}$ for the two-dimensional problem can be obtained by differentiation of (A1.4) and (A1.5), the resulting kernels again being nonintegrable. Its discrete counterpart $S_{\alpha\beta}$ can be obtained using (5.2.12).

The discrete Green functions $F_{\alpha\beta}$ and $S_{\alpha\beta}$ for three-dimensional problems were determined by Das (1980) and by Das and Kostrov (1987), respectively. The discrete F_{11} for the inplane shear problem was determined following Andrews (1985) and by carrying out the integrations analytically. The F_{22} for the antiplane case given above was determined by Andrews (personal communication). Burridge (1969) developed a method of obtaining discrete kernels similar to $S_{\alpha\beta}$ without the use of analytic Green functions and applied the method to some two-dimensional problems. We will not discuss Burridge's method here except to note that the resulting discrete kernel had some oscillations along the time axis similar to what is found for $S_{\alpha\beta}$ in the three-dimensional case. Burridge and Moon (1981) applied the method of Burridge (1969) to a three-dimensional scalar problem.

App. 2 Approximation and stability properties of the representation relations

We first discuss the approximation error introduced into the representation relation (5.1.9) by replacing the stress components and Green functions by their grid-averaged values (5.2.2) and (5.2.4). As mentioned in Section 5.6, the singular behavior of a perfectly brittle body is only a mathematical model, and for our purposes can be considered a limit of the nonideally brittle body model with a smooth distribution of stress. The approximation and stability of the problem with a singular stress can be examined, but this will require a more sophisticated mathematical treatment than we consider here and would not be appropriate in this book. We shall therefore study the approximation error assuming that the traction perturbation τ_α is finite everywhere on the fault plane $X_3 = 0$. Then, the value of τ_α at any point within the (i, j, k) element of the numerical scheme is

$$\tau_\alpha(\mathbf{X}, t) = \tau_{\alpha ijk} + o(1) \tag{A2.1}$$

when $\Delta X \to 0$ and $\Delta t / \Delta X$ is fixed. The approximation error can be estimated only for a particular choice of boundary conditions (5.7.1) but is usually (and this is true for all the examples of boundary conditions considered in Chapter 5) of the order $(\Delta X)^\lambda$, $\lambda > 0$.

The slip a_α [equation (5.2.3)] at any point within the grid is then given by equation (5.2.8) as

$$a_{\alpha ijk} = - \sum_{k'=0}^{k} \sum_{i'=-\infty}^{\infty} \sum_{j'=-\infty}^{\infty} F_{\alpha\beta}(i - i', j - j', k - k') \tau_{\alpha i' j' k'}$$

$$+ \text{approximation error} \tag{A2.2}$$

The summations in the analytic form of the representation relation (5.1.9) extend over the P cone. However, because of the finiteness of the grids, some grids centered outside the P cone intersect it, and the corresponding elements of $F_{\alpha\beta}$ are nonvanishing. As a result of this, the above expression produces disturbances that propagate at speeds greater than v_P for some choices of $v_P \Delta t / \Delta X$. This problem can be avoided in the following manner. Expression (A2.2) is not the only possible approximation of integral (5.1.9). A few additional terms can be added to it, provided that their sum tends toward zero as $\Delta X \to 0$. Now from condition (A2.1), the a_α's are smooth functions, so that the difference

$$a_{\alpha i\,jk} - a_{\alpha i\,j(k-1)}$$

is of order ΔX. Consequently, if k on the right side of (A2.2) is replaced by $k - n$, where n is independent of the grid size, the order of approximation will remain unchanged. Next, adding n terms of the form $\Delta t F_0 \tau_{\alpha i\,jk}$, where $k - n < k' < k$, we obtain, instead of (A2.2), with the same order of accuracy of approximation the expression

$$a_{\alpha i\,jk} = - \sum_{k'=0}^{k} \sum_{i'=-\infty}^{\infty} \sum_{j'=-\infty}^{\infty} F_{\alpha\beta}^*(i - i',\, j - j',\, k - k') \tau_{\beta i'\,j'k'}$$

where

$$F_{\alpha\beta}^*(i,\, j,\, k) = \begin{cases} F_{\alpha\beta}(i,\, j,\, k - n) & \text{for} \quad k > n \\ \Delta t F_0 \delta_{\alpha\beta} & \text{for} \quad 0 < k \leq n \end{cases} \qquad \text{(A2.3)}$$

Selecting sufficiently large n, we can thus shift all nonvanishing elements of $F_{\alpha\beta}^*$ into the P cone. The minimum value of n necessary for this obviously depends on the ratio $v_P \Delta t/\Delta X$. Note that the tip value $F_{\alpha\beta}^*(0, 0, 0)$ differs from (5.2.6) in that it contains the factor Δt instead of δt. For $\delta t < \Delta t$, this difference is of the order Δt, which is negligible from the point of view of accuracy of the approximation. In general, this element can be multiplied by any constant factor, a possibility that will be important for an analysis of the stability of the algorithm. The above discussion implies that whatever the disturbances that may appear outside the P cone when approximation (A2.2) is used, they will be of the order of the approximation error. Consequently, the replacement of $F_{\alpha\beta}$ by $F_{\alpha\beta}^*$ is optional. In the rest of this appendix, we shall use $F_{\alpha\beta}$ to denote either of these two quantities.

A detailed investigation of the stability criterion for this algorithm is a task well beyond the scope of this book. Here, we shall consider only a necessary condition, which is that the algorithm be stable when applied to very simple problems. Let us consider the problem in which the known slip on $X_3 = 0$ is proportional to time and independent of X_1 and X_2, that is,

$$\left.\begin{array}{l} a_1 = At \\ a_2 = 0 \end{array}\right\} \quad \text{for} \quad X_3 = 0,\, -\infty < X_1,\, X_2 < \infty \qquad \text{(A2.4)}$$

This problem has a trivial solution in the form of a plane shear wave

$$a_\alpha(\mathbf{X}, t) = \delta_{\alpha 1} \cdot A \cdot \left(t - \frac{X_3}{v_S}\right) \tag{A2.5}$$

where v_S is the shear wave velocity. The tractions at $X_3 = 0$ will be

$$\tau_\alpha = -\mu \frac{A}{v_S} \qquad \text{at} \quad X_3 = 0 \tag{A2.6}$$

independent of X_1, X_2, and t. Now assign U_α, defined by (5.7.2) as

$$U_\alpha = \delta_{\alpha 1} \cdot A \cdot (k \, \Delta t + \delta t)$$

and apply algorithm (5.2.9) together with the boundary condition (5.7.1). Obviously, $a_{\alpha i j k}$ and $\tau_{\alpha i j k}$ will not depend on i and j. Let us denote them by $a_{\alpha k}$ and $\tau_{\alpha k}$. From (5.7.1), we have

$$a_{\alpha k} = \delta_{\alpha 1} \cdot A \cdot (k \, \Delta t + \delta t)$$

and substituting it into (5.2.9), we obtain

$$\tau_{\alpha k} = -\frac{F_0}{G_0} \sum_{k'=0}^{k-1} \tau_{\alpha k'} - \frac{\delta_{\alpha 1} \cdot A \cdot (k \, \Delta t + \delta t)}{G_0 \, \Delta t} \tag{A2.7}$$

where the property (A1.3) is used. (Note that G_0 is a parameter to be chosen from stability considerations.) For $k = 0$ and $k = 1$, we get

$$\tau_{\alpha 0} = -\frac{\delta_{\alpha 1} \cdot \delta t \cdot A}{G_0 \, \Delta t} \tag{A2.8}$$

$$\tau_{\alpha 1} = -\frac{\delta_{\alpha 1} \cdot A}{G_0} \left(1 + \frac{\delta t}{\Delta t}\left(1 - \frac{F_0}{G_0}\right)\right) \tag{A2.9}$$

For $k > 0$, we subtract from (A2.7) the equation obtained from it by substituting $k - 1$ instead of k, and obtain

$$\tau_{\alpha k} - \left(1 - \frac{F_0}{G_0}\right) \tau_{\alpha(k-1)} = -\delta_{\alpha 1} \frac{A}{G_0}$$

which is a linear finite difference equation of first order with constant coefficients. Solving it with the initial condition (A2.8), we get

$$\tau_{\alpha k} = -\frac{\delta_{\alpha 1} A}{F_0}\left[1 - \left(1 - \frac{\delta t F_0}{\Delta t G_0}\right)\left(1 - \frac{F_0}{G_0}\right)^k\right] \qquad \text{for} \quad k \geq 1$$

This solution is unstable (grows exponentially) if

$$\left| 1 - \frac{F_0}{G_0} \right| > 1$$

which gives the necessary condition for stability for this problem in the form

$$0 < \frac{F_0}{G_0} < 2$$

This solution is most stable (constant) when G_0 is chosen equal to F_0. In this case, we have

$$\tau_{\alpha 0} = - \frac{\delta_{\alpha 1} \cdot A \cdot \delta t}{F_0 \, \Delta t}$$

$$\tau_{\alpha k} = - \frac{\delta_{\alpha 1} \cdot A}{F_0} \qquad \text{for} \quad k \geq 1$$

(A2.10)

Substituting the exact solution (A2.5) and (A2.6) into (5.1.9) and using the definition of F_0 from Appendix 1, we obtain

$$F_0 = v_{\rm S}/\mu$$

(A2.11)

which is a positive constant, as expected from Appendix 1. The solution (A2.10) coincides with the exact solution (A2.6) for $k \geq 1$.

For all the two- and three-dimensional shear crack problems for which the Green functions were given in Appendix 1, it was found that F_0 was equal to $v_{\rm S}/\mu$. (The above stability criterion was used in all the applications of this method in Chapter 5.) The results showed appreciable stability and agreed well with available analytic solutions (Section 3.5). The stability criterion and approximation error considerations for algorithm (5.2.13) are the same as those for algorithm (5.1.9) discussed here. Similar considerations can be undertaken for a tension crack problem following the discussion in this appendix, if so desired.

REFERENCES

Achenbach, J. D., and Abo-Zena, A. M. (1973). Analysis of the dynamics of strike-slip faulting, *J. Geophys. Res. 78*, 866–75.

Aggarwal, Y. P., Sykes, L. R., Armbruster, J., and Sbar, M. L. (1973). Premonitory changes in seismic velocities and prediction of earthquakes, *Nature (London) 241*, 101–4.

Aki, K. (1966). Generation and propagation of G-waves from the Niigata earthquake of June 16, 1964. 2. Estimation of earthquake movement, released energy and stress–strain drop from G-wave spectrum, *Bull. Earthquake Res. Inst. 44*, 23–88.

(1967). Scaling law of seismic spectrum, *J. Geophys. Res. 72*, 1217–31.

(1972a). Earthquake mechanism, in *The Upper Mantle*, ed. A. R. Ritsema (New York: Elsevier), 423–46.

(1972b). Scaling law of earthquake source time-function, *Geophys. J. Roy. Astr. Soc. 31*, 3–25.

(1979). Characterization of barriers on an earthquake fault, *J. Geophys. Res. 84*, 6140–8.

Aki, K., and Richards, P. G. (1980). *Quantitative Seismology: Theory and Methods* (San Francisco: Freeman).

Andrews, D. J. (1976a). Rupture velocity of plane strain shear cracks, *J. Geophys. Res. 81*, 5679–87.

(1976b). Rupture propagation with finite stress antiplane strain, *J. Geophys. Res. 81*, 3575–82.

(1980). A stochastic fault model. I. Static case, *J. Geophys. Res. 85*, 3867–77.

(1982). A stochastic fault model. II. Time-dependent case, *J. Geophys. Res. 86*, 10821–34.

(1985). Dynamic plane-strain shear rupture with a slip-weakening friction law calculated by a boundary integral method, *Bull. Seismol. Soc. Am. 75*, 1–22.

Archuleta, R. J. (1976). "Experimental and Numerical Three Dimensional Simulations of Strike-Slip Earthquakes" (Ph.D. dissertation, University of California, San Diego).

(1984). Faulting model for the 1979 Imperial Valley earthquake, *J. Geophys. Res. 89*, 4559–85.

Archuleta, R. J., and Frazier, G. A. (1978). Three dimensional numerical simulations of dynamic faulting in a half-space, *Bull. Seismol. Soc. Am. 68*, 573–98.

Atkinson, B. K. (1979). A fracture mechanics study of subcritical tensile cracking of quartz in wet environments, *Pageoph 117*, 1011–24.

Atluri, S. N., and Nishioka, T. (1985). Numerical studies in dynamic fracture mechanics, *Intl. J. Frac. 27*, 245–61.

Backus, G. E., and Mulcahy, M. (1976a). Moment tensors and other phenomenological descriptions of seismic sources. I. Continuous displacements, *Geophys. J. Roy. Astr. Soc. 46*, 341–61.

Backus, G. E., and Mulcahy, M. (1976b). Moment tensors and other phenomenological descriptions of seismic sources. II. Discontinuous displacements, *Geophys. J. Roy. Astr. Soc. 47*, 301–29.

Baker, B. R. (1962). Dynamic stresses created by a moving crack, *J. Appl. Mech. 29*, 449–58.

Barenblatt, G. I. (1959). The formation of equilibrium cracks during brittle fracture. General ideas and hypotheses, *J. Appl. Math. Mech., 23*, 622–36.

Barka, A. A., and Kadinsky-Cade, K. (1988). Strike slip fault geometry in Turkey and its influence on earthquake activity, *Tectonics 7*, 663–84.

Benioff, H. (1951). Earthquakes and rock creep. Part I. Creep characteristics of rocks and the origin of aftershocks, *Bull. Seismol. Soc. Am. 41*, 31–62.

(1955). Mechanism and strain characteristics of the White Wolf fault as indicated by the aftershock sequence, *Earthquakes in Kern County, California during 1952*, Bull. 171, Div. Mines, State of California, November, 199–202.

Ben-Menahem, A., and Toksöz, M. N. (1963). Source mechanism from spectra of long period surface waves, *J. Geophys. Res. 68*, 5207–22.

Berg, C. A. (1967). A note on the mechanics of seismic faulting, *Geophys. J. Roy. Astr. Soc. 14*, 89–99.

Boatwright, J. L. (1980). A spectral theory for the circular seismic sources: Simple estimates of source dimension, dynamic stress-drop and radiated seismic energy, *Bull. Seismol. Soc. Am. 70*, 1–27.

(1981). Quasi-dynamic models of simple earthquakes: An application to an aftershock of the 1975 Oroville, California earthquake, *Bull. Seismol. Soc. Am. 71*, 69–94.

(1984). Seismic estimates of stress release, *J. Geophys. Res. 89*, 6961–8.

Boore, D. M., McEvilly T. V., and Lindh, A. (1975). Quarry blast sources and earthquake prediction: The Parkfield, California earthquake of June 28, 1966, *Pageoph 113*, 293–6.

Broberg, K. B. (1960). The propagation of a brittle crack, *Arch. Fys. 18*, 159–92.

Bruhn, R. L., Gibler, P. R., and Parry, W. T. (1987). Rupture characteristics of normal faults: An example from the Wasatch fault zone, Utah, *Continental Extensional Tectonics*, Geol. Soc. Special Publ., *29*, 337–53.

Brune, J. N. (1968). Seismic moment, seismicity and rate of slip along major fault zones, *J. Geophys. Res. 73*, 777–84.

Brune, J. N., Archuleta, R. J., and Hartzell, S. (1979). Far-field S-wave spectra, corner frequencies and pulse shapes, *J. Geophys. Res. 84*, 2262–72.

Budiansky, B., and Rice, J. R. (1979). An integral equation for dynamic elastic response of an isolated 3-D crack, *Wave Motion 1*, 187–92.

Bullen, K. E. (1953). On strain energy and strength in the earth's upper mantle, *Trans. Am. Geophys. Union 34*, 107–9.

(1963). *An Introduction to the Theory of Seismology* (Cambridge University Press, 1963).

Burridge, R. (1969). The numerical solution of certain integral equations with non-integrable kernels arising in the theory of crack propagation and elastic wave diffraction, *Phil. Trans. Roy. Soc. London A265*, 353—81.

(1973). Admissible speeds for plane-strain self-similar shear crack with friction but lacking cohesion, *Geophys. J. Roy. Astr. Soc. 35*, 439–55.

Burridge, R., Conn, G., and Freund, L. B. (1979). The stability of a rapid Mode II shear crack with finite cohesive traction, *J. Geophys. Res. 84*, 2210–22.

Burridge, R., and Knopoff, L. (1964). Body force equivalents for seismic dislocations, *Bull. Seismol. Soc. Am. 54*, 1875–88.

Burridge, R., and Moon, R. (1981). Slipping on a frictional fault plane in three dimensions: A numerical simulation of a scalar analog, *Geophys. J. Roy. Astr. Soc. 67*(2), 325–42.

Burridge, R., and Willis, J. R. (1969). The self-similar problem of the expanding elliptical crack in an anisotropic solid, *Proc. Camb. Phil. Soc. 66*, 443–68.

Byerlee, J. D. (1967). Frictional characteristics of granite under high confining pressure, *J. Geophys. Res. 72*, 3639–48.

Chao, C. C. (1960). Dynamical response of an elastic half-space to tangential surface loadings, *J. Appl. Mech. 27*, 559–67.

Cherepanov, G. P. (1967). Crack propagation in continuous media, *J. Appl. Math. Mech.* (English transl.) *31*, 503–12.

(1979). *Mechanics of Brittle Fracture* (New York: McGraw-Hill).

Chinnery, M. A. (1969). Theoretical fault models, *Publ. Dominion Observ. Ottawa 37*, 211–23.

Chouet, B. Aki, K., and Tsujiura, M. (1978). Regional variation of the scaling law of earthquake source spectra, *Bull. Seismol. Soc. Am. 68*, 49–79.

Cichowicz, A. (1981). Determination of source parameters from seismograms of mining tremors and the inverse problem for a seismic source, *Publ. Inst. Geophys.*, Polish Acad. Sci. *A-11*(147).

Cox, S. J. D., and Atkinson, B. K. (1983). Fracture mechanics and acoustic emission of antiplane shear cracks in rocks, *Earthquake Predict. Res. 2*, 1–23.

Cox, S. J. D., and Scholz, C. H. (1985). A direct measurement of shear fracture energy in rocks, *Geophys. Res. Lett. 12*, 813–16.

Das, S. (1976). "A Numerical Study of Rupture Propagation and Earthquake Source Mechanism" (Sc.D. thesis, Massachusetts Institute of Technology, Cambridge).

(1980). A numerical method for determination of source time functions for general three-dimensional rupture propagation, *Geophys. J. Roy. Astr. Soc. 62*, 591–604.

(1981). Three-dimensional rupture propagation and implications for the earthquake source mechanism, *Geophys. J. Roy. Astr. Soc. 67*, 375–93.

(1982). Appropriate boundary conditions for modelling very long earthquakes and physical consequences, *Bull. Seismol. Soc. Am. 72*, 1911–26.

(1985). Application of dynamic shear crack models to the study of the earthquake faulting process, *Intl. J. Frac. 27*, 263–76.

Das, S., and Aki, K. (1977a). A numerical study of two-dimensional rupture propagation, *Geophys. J. Roy. Astr. Soc. 50*, 643–68.

(1977b). Fault plane with barriers: A versatile earthquake model, *J. Geophys. Res. 82*, 5658–70.

Das, S., Boatwright, J., and Scholz, C. H., Eds. (1986). *Earthquake Source Mechanics*, Fifth Maurice Ewing Symposium Proceedings, Vol. 6, American Geophysical Union Monograph No. 37 (Washington, D.C.).

Das, S., and Kostrov, B.V. (1983). Breaking of a single asperity: Rupture process and seismic radiation, *J. Geophys. Res. 88*, 4277–88.

(1985). An elliptical asperity in shear: Fracture process and seismic radiation, *Geophys. J. Roy. Astr. Soc. 80*, 725–42.

(1986). Fracture of a single asperity on a finite fault: A model for weak earthquakes? in *Earthquake Source Mechanics*, ed. S. Das, J. Boatwright, and C. H. Scholz, American Geophysical Union Monograph No. 37 (Washington, D.C.), 91–6.

(1987). On the numerical boundary integral equation method for three-dimensional dynamic shear crack problems. *J. Appl. Mech. 54*, 99–104.

Day, S. M. (1982a). Three-dimensional simulation of spontaneous rupture: The effect of non-uniform prestress, *Bull. Seismol. Soc. Am. 72*, 1881–1902.

(1982b). Three-dimensional finite difference simulation of fault dynamics: Rectangular faults with fixed rupture velocity, *Bull. Seismol. Soc. Am. 72*, 705–27.

Dieterich, J. H. (1978). Time-dependent friction and the mechanics of stick-slip, *Pageoph 116*, 790–806.

(1979). Modeling of rock friction. 1. Experimental results and constitutive equations, *J. Geophys. Res. 84*, 2161–8.

Dolbilkina, N. A., Myachkin, V. I., Palyonov, A. M., and Preobrazenskii, V. B. (1979). The study of time variations of elastic-wave parameters: Results and problems. *Phys. Earth Planet. Interiors 18*, 319–25.

Dugdale, D. S. (1960). Yielding of steel sheets containing slits. *J. Mech. Phys. Solids 8*, 100–10.

Dziewonski, A. M., and Woodhouse, J. H. (1983). Studies of the seismic source using normal-mode theory, in *Earthquakes: Observation, Theory, and Interpretation*, ed. H. Kanamori and E. Boschii (Amsterdam: North-Holland).

Ekstrom, G. A. (1987). "A Broadband Method of Earthquake Analysis" (Ph.D. Thesis, Harvard University).

Eshelby, J. D. (1957). The determination of the elastic field of an ellipsoidal inclusion, and related problems, *Proc. Roy. Soc. Lond. A241*, 376–96.

(1969). The elastic field of a crack extending non-uniformly under general antiplane loading, *J. Mech. Phys. Solids 17*, 177–99.

Freund, L. B. (1972a). Crack propagation in an elastic solid subjected to general loading. I. Constant rate of extension, *J. Mech. Phys. Solids 20*, 129–40.

(1972b). Crack propagation in an elastic solid subjected to general loading. II. Nonuniform rate of extension, *J. Mech. Phys. Solids 20*, 141–52.

(1972c). Energy flux into the tip of an extending crack in an elastic solid, *J. Elasticity 2*, 341–9.

(1972d). The initial wavefront emitted by a suddenly extending crack in an elastic solid, *J. Appl. Mech. 39*, 601–2.

(1973). Crack propagation in an elastic solid subjected to general loading. III. Stress wave loading, *J. Mech. Phys. Solids 21*, 47–61.

(1974). Crack propagation in an elastic solid subjected to general loading. IV. Obliquely incident stress pulse, *J. Mech. Phys. Solids 22*, 137–46.

(1979). The mechanics of dynamic shear crack propagation, *J. Geophys. Res. 84,* 2199–2209.

Green, A. E., and Zerna, W. (1954). *Theoretical Elasticity* (New York: Oxford University Press).

Griffith, A. A. (1921). The phenomenon of rupture and flow in solids, *Phil. Trans. Roy. Soc. London, Ser. A. 221,* 163–98.

Gutenberg, B. (1955). Magnitude determination for larger Kern County shocks, 1952: Effects of station azimuth and calculation methods, in *Earthquakes in Kern County, California during 1952,* Bull. 171, Div. Mines, State of California, November, 171–6.

Hadley, K. (1975). V_p/V_s Anomalies in dilatant rock samples, *Pageoph 113,* 1–23.

(1976). The effect of cyclic stress on dilatancy: Another look, *J. Geophys. Res. 81,* 2471–4.

Hamano, Y. (1974). Dependence of rupture time history on the heterogeneous distribution of stress and strength on the fault (abstract), *EOS, Trans. Am. Geophys. Union 55,* 352.

Hanks, T. C. (1976). Observations and estimation of long-period strong ground motion in the Los Angeles basin, *Earthquake Eng. Struct. Dynam. 4,* 473–88.

Hartzell, S. H. (1978). Earthquake aftershocks as Green's functions, *Geophys. Res. Lett. 5,* 1–4.

Hartzell, S. H., and Heaton, T. H. (1983). Inversion of strong-ground motion and teleseismic waveform data for the fault rupture history of the 1979 Imperial Valley, California earthquake, *Bull. Seismol. Soc. Am. 73,* 1553–83.

(1986). Rupture history of the 1984 Morgan Hill, California earthquake from the inversion of strong motion records, *Bull. Seismol. Soc. Am. 76,* 649–74.

Haskell, N. A. (1964). Total energy and energy spectral density of elastic wave radiation from propagating faults, *Bull. Seismol. Soc. Am. 54,* 1811–41.

Helmberger, D. V. (1983). Theory and application of synthetic seismograms, in *Earthquakes: Observation, Theory, and Interpretation,* ed. H. Kanamori and E. Boschii, (Amsterdam: North-Holland).

Herglotz, G. (1907). Ueber das Benndorfsche Problem der Fortpflanzungsgeschwinigkeit der Erdebeben strahlen, *Phys. Z. 8,* 145–7.

Hodgson, J. H., Ed. (1959). The Mechanics of Faulting, with Special Reference to the Fault Plane Work (symposium), *Publ. Dominion Observ. Ottawa 20*(2).

(1960). A Symposium on Earthquake Mechanism (Helsinki, 1960), *Publ. Dominion Observ. Ottawa 24*(10).

House, L., and Boatwright, J. L. (1980). Investigation of two high stress-drop earthquakes in the Shumagin seismic gap, Alaska, *J. Geophys. Res. 85,* 7151–65.

Husseini, M. I., Jovanovich, D. B., Randall, M. J., and Freund, L. B. (1975). The fracture energy of earthquakes, *Geophys. J. Roy. Astr. Soc. 43,* 367–85.

Ida, Y. (1972). Cohesive force across the tip of a longitudinal-shear crack and Griffith's specific surface energy, *J. Geophys. Res. 77,* 3796–3805.

(1973). The maximum acceleration of seismic ground motion, *Bull. Seismol. Soc. Am. 63,* 959–68.

Irwin, G. R. (1948). Fracture dynamics, in *Fracturing of Metals* (Cleveland, Ohio: American Society for Metals), 147–66.

(1969). Basic concepts for dynamic fracture testing, *Trans. ASME 91,* 519–24.

Johnson, T. (1981). Time-dependent friction of granite: Implications for precursory slip on faults, *J. Geophys. Res. 86,* 6017–28.

Johnson, T., and Scholz, C. H. (1976). Dynamic properties of stick-slip friction of rocks, *J. Geophys. Res. 81,* 881–8.

Joyner, W. B., and Boore, D. M. (1986). On simulating large earthquakes by Green's function addition of smaller earthquakes, in *Earthquake Source Mechanics*, ed. S. Das, J. Boatwright, and C. H. Scholz, American Geophysical Union Monograph No. 37 (Washington, D.C.), 269–74.

Kanamori, H., and Anderson, D. L. (1975). Theoretical basis of some empirical relations in seismology, *Bull. Seismol. Soc. Am. 65,* 1073–95.

Kanamori, H., and Cipar, J. J. (1974). Focal process of the great Chilean earthquake May 22, 1960, *Phys. Earth Planet. Interiors 9,* 128–36.

Kanamori, H., and Stewart, G. S. (1978). Seismological aspects of the Guatemala earthquake, *J. Geophys. Res. 83,* 3427–34.

Kanninen, M. F., and Popelar, C. H. (1985). *Advanced Fracture Mechanics* (New York: Oxford University Press).

Kikuchi, M., and Kanamori, H. (1982). Inversion of the complex body waves, *Bull. Seismol. Soc. Am. 72,* 491–506.

King, G., and Yielding, F. (1984). The evolution of a thrust fault system: Processes of rupture initiation, propagation and termination in the 1980 El Asnam (Algeria) earthquake, *Geophys. J. Roy. Astr. Soc. 77,* 915–33.

Knopoff, L. (1958). Energy release in earthquakes, *Geophys. J. Roy. Astr. Soc. 7,* 44–52.

Kostrov, B. V. (1964). Self similar problems of propagation of shear cracks, *J. Appl. Math. Mech. 28,* 1077–87.

  (1966). Unsteady propagation of longitudinal shear cracks, *J. Appl. Math. Mech. 30,* 1241–8.

  (1974). Seismic moment and energy of earthquakes and seismic flow of rock, *Izv., Phys. Solid Earth. 13,* 13–21.

Kostrov, B. V., and Das, S. (1984). Evaluation of stress and displacement fields due to an elliptical plane shear crack, *Geophys. J. Roy. Astr. Soc. 78,* 19–33.

Kostrov, B. V., Nikitin, L. V., and Flitman, L. M. (1969). The mechanics of brittle fracture, *MTT* (English transl.) *4,* 112–25.

Kuznetsova, K. I., Aptekman, Z. Y., Shebalin, N. V., and Shteynberg, V. V. (1976). Aftershocks of relaxation and aftershocks due to fracture growth of the Dagestan earthquake, in *Investigations in Earthquake Physics* (in Russian) (Moscow: Nauka), 94–113.

Lamb, H. (1904). On the propagation of tremors at the surface of an elastic solid, *Phil. Trans. Roy. Soc. Lond. Ser. A, 203,* 1–42.

Leonov, M. Y., and Panasyuk, V. V. (1959). Development of the most dangerous cracks in a solid, *Prikladna Mekhanika* (in Ukrainian), no. 5.

Lindh, A. G., and Boore, D. M. (1981). Control of rupture by fault geometry during the 1966 Parkfield earthquake, *Bull. Seismol. Soc. Am. 71,* 95–116.

Love, A. E. H. (1944). *A Treatise on the Mathematical Theory of Elasticity* (New York: Dover).

Madariaga, R. (1976). Dynamics of an expanding circular fault, *Bull. Seismol. Soc. Am. 66,* 639–66.

  (1981). Dynamics of seismic sources, in *Identification of Seismic Sources: Earthquakes or Underground Explosions,* ed. E. S. Husebye and S. Mykkeltveit (Dordrecht: Reidel), 71–96.

Magistrale, H., and Kanamori, H. (1986). Changes in NTS seismograms recorded in the source region of the North Palm Springs earthquake of 7/8/86. *Trans. Am. Geophys. Union 67,* 1090.

Mandelbrot, B. (1977). *Fractals* (San Francisco: Freeman).

Martynov, V. G. (1983). Spectra of S-waves for small local earthquakes: Possibilities for their use in seismic prediction, in *Experimental Seismology* (in Russian) (Moscow: Nauka), 128–43.

McClintock, F. A., and Argon, A. S. (1966). *Mechanical Behavior of Materials* (Reading, Mass.: Addison-Wesley).

McGarr, A., Spottiswoode, S. M., Gay, N. C., and Ortlepp, W. D. (1979). Observations relevant to seismic driving stress, stress drop and efficiency, *J. Geophys. Res. 84,* 2251–61.

Mikumo, T., and Miyatake, T. (1979). Earthquake sequences on frictional fault model with a non-uniform strength and relaxation times, *Geophys. J. Roy. Astr. Soc. 59,* 497–522.

Mindlin, R. D. (1949). Compliance of elastic bodies in contact, *J. Appl. Mech. 16,* 259–68.

Molnar, P., Tucker, B., and Brune, J. N. (1973). Corner frequencies of P and S waves and models of earthquake sources, *Bull. Seismol. Soc. Am. 65,* 2091–2104.

Myachkin, V. I. (1978). *Earthquake Preparation Processes* (in Russian) (Moscow: Nauka).

Myachkin, V. I., Dolbilkina, N. A., Kushnir, G. S., Levshenko, V. T., Maksimov, O. A., Paleonov, A. M., Preobrazhenskii, V. B., and Solov'eva, R. P. (1985a). Some observational data on sounding focal zones in Kamchatka and their accuracy, in *Physics of the Earthquake Focus* (Moscow: Nauka, 1975), 174–86. (English translation for U.S. Dept. of the Interior and National Science Foundation, Washington, D.C., by Amerind Publ., New Delhi, 1985.)

Myachkin, V. I., Kostrov, B. V., Sobolev, G. A., and Shamina, O. G. (1985b). Fundamentals of physics of the earthquake focus forerunners, in *Physics of the Earthquake Focus* (Moscow: Nauka, 1985b), 1–24. (English translation for U.S. Dept. of the Interior and National Science Foundation, by Amerind Publ., New Delhi, 1985.)

Myachkin, V. I., Sobolev, G. A., Dolbilkina, N. A., Morozow, V. N., and Preobrazhenskii, V. B. (1972). Study of variations in geophysical fields near focal zones of Kamchatka, *Tectonophysics 14,* 287–93.

Nabelek, J., and King, G. (1985). Role of fault bends in the initiation and termination of earthquake rupture, *Science 228,* 984–7.

Nakano, H. (1923). Notes on the nature of forces which give rise to earthquake motions, *Seismol. Bull., Cent. Meteor. Obs. Japan 1,* 92–120.

Nersesov, I. L., Semenov, A. N., and Simbireva, I. G. (1971). Space–time distribution of t_s/t_p in Garm, in *Experimental Seismology* (in Russian) (Moscow: Akad. Nauka SSR Publ.).

Nowacki, W. (1963). *Dynamics of Elastic Systems* (New York: Wiley).

Nur, A. (1972). Dilatancy, pore fluids and premonitory variations of t_s/t_p travel times, *Bull. Seismol. Soc. Am. 62,* 1217–22.

Okubo, P. G. (1986). "Experimental and Numerical Model Studies of Frictional Instability Seismic Sources" (Ph.D. thesis, Massachusetts Institute of Technology, Cambridge).

Okubo, P. G., and Dieterich, J. H. (1981). Fracture energy of stick-slip events in a large scale biaxial experiment, *Geophys. Res. Lett. 8,* 887–90.

Olson, A. H., and Apsel, R. J. (1982). Finite faults and inverse theory with applications to the 1979 Imperial Valley earthquakes, *Bull. Seismol. Soc. Am. 72,* 1969–2001.

Orowan, E. (1952). Fundamentals of brittle behavior in metals, in *Fatigue and Fracture of Metals,* ed. William M. Murray (New York: Wiley), 139–62.

Palmer, A. C., and Rice, J. R. (1973). The growth of slip surfaces in the progressive failure of over-consolidated clay, *Proc. Roy. Soc. Lond. A. 332*, 527–48.

Randall, M. J. (1971). Elastic multipole theory and seismic moment, *Bull. Seismol. Soc. Am. 61*, 1321–6.

Reid, H. F. (1910). The mechanics of the earthquake, in *The California Earthquake of April 18, 1906*, Report of the State Investigation Commission, Vol. 2 (Washington, D.C.: Carnegie Institute of Washington).

Rice, J. R. (1968). A path independent integral and the approximate analysis of strain concentration by notches and cracks, *J. Appl. Mech. 35*, 379–86.

(1980). The mechanics of earthquake rupture, in *Physics of the Earth's Interior* (Bologna: Soc. Italiana di Fisica), 555–649.

Rice, J. R., and Simons, D. A. (1976). The stabilization of spreading shear faults by coupled deformation–diffusion effects in fluid infiltrated porous materials, *J. Geophys. Res. 81*, 5322–34.

Richards, P. G. (1979). Elementary solutions to Lamb's problem for a point source and their relevance to three-dimensional studies of spontaneous crack propagation, *Bull. Seismol. Soc. Am. 69*, 947–56.

Riznichenko, Y. V. (1965). Seismic rock flow, in *Dynamics of the Earth's Crust* (in Russian) (Moscow: Nauka).

Rudnicki, J. W. (1980). Fracture mechanics applied to the earth's crust, *Annu. Rev. Earth Planet. Sci. 8*, 489–525.

Rudnicki, J. W., and Freund, L. B. (1981). On energy radiation from seismic sources, *Bull. Seismol. Soc. Am. 71*, 583–95.

Rudnicki, J. W., and Kanamori, H. (1981). Effects of fault interaction on moment, stress-drop and strain energy release, *J. Geophys. Res. 86*, 1785–93.

Ruff, L. J. (1983). Fault asperities inferred from seismic body waves, in *Earthquakes: Observation, Theory and Interpretation*, ed. H. Kanamori and E. Boschii (Amsterdam: North-Holland), 251–76.

Ruff, L. J., and Kanamori, H. (1983). The rupture process and asperity distribution of three great earthquakes from long-period diffracted P-waves, *Phys. Earth Planet. Interiors 31*, 202–30.

Ruina, A. (1980). Friction laws and instabilities: A quasistatic analysis of some dry frictional behavior, Rep. No. 21, Div. Engineering (Providence, R.I.: Brown University).

(1983). Slip instability and state variable friction laws, *J. Geophys. Res. 88*, 10359–70.

Rummel, F., Alheid, H. J., and Frohn, C. (1978). Dilatancy and fracture induced velocity changes in rock and their relation to frictional sliding, *Pageoph 116*, 743–64.

Sato, T., and Hirasawa, T. (1973). Body wave spectra from propagating shear cracks, *J. Phys. Earth 21*, 415–31.

Scholz, C. H. (1982). Scaling laws for large earthquakes: Consequences for physical models, *Bull. Seismol. Soc. Am. 72*, 1–14.

Scholz, C. H., Molnar, P. M., and Johnson, T. (1972). Detailed studies of frictional sliding of granite and implications for the earthquake mechanism. *J. Geophys. Res. 77*, 6392–6406.

Semenov, A. N. (1969). Variations in the travel time of transverse and longitudinal waves before a violent earthquake, *Bull. Acad. Sci. USSR, Phys. Solid Earth 3*, 245–8.

Sibson, R. (1986). Rupture interaction with fault jogs, in *Earthquake Source Mechanics*, ed. S. Das, J. Boatwright, and C. H. Scholz, American Geophysical Union Monograph No. 37 (Washington, D.C.), 157–67.

Slepyan, L. I. (1976). Crack dynamics in an elastic-plastic body, *MTT* (English transl.) *11*, 126–34.

Sobolev, G., and Rummel, F. (1982). Shear fracture development and seismic regime in pyrophyllite specimens with soft inclusions, *J. Geophys. 51*, 180–7.

Spottiswoode, S. M., and McGarr, A. (1975). Source parameters of tremors in a deep-level gold mine, *Bull. Seismol. Soc. Am. 65*, 93–112.

Takeuchi, H., and Kikuchi, M. (1973). A dynamical model of crack propagation, *J. Phys. Earth 21*, 27–37.

Titchmarsh, E. C. (1948). *Theory of Fourier Integrals* (New York: Oxford University Press).

Truesdell, C. (1965–6). *Continuum Mechanics.* Part I, *The Mechanical Foundations of Elasticity and Fluid Dynamics.* Part II, *The Rational Mechanics of Materials.* Part III, *Foundations of Elasticity Theory* (New York: Gordon & Breach).

Tsuboi, C. (1956). Earthquake energy, earthquake volume, aftershock area and strength of the earth's crust, *J. Phys. Earth 4*, 63–6.

Virieux, J., and Madariaga, R. (1982). Dynamic faulting studied by a finite difference method, *Bull. Seismol. Soc. Am. 72*, 345–69.

Vvedenskaya, A. V. (1956). The determination of displacement fields by means of dislocation theory, *Izv. Akad. Nauka SSSR, Ser. Geofiz*, no. 3, 227–84.

Walsh, J. B. (1968). Mechanics of strike-slip faulting with friction, *J. Geophys. Res. 73*, 761–76.

Wawersik, S. R., and Brace, W. F. (1971). Post-failure behavior of a granite and a diabase, *Rock Mech. 3*, 61–85.

Wesnousky, S. G., Scholz, C. H., and Shimazaki, K. (1982). Deformation of an island arc: Rates of moment release and crustal shortening in intraplate Japan determined from seismicity and quaternary fault data, *J. Geophys. Res. 87*, 6829–52.

Whitcomb, J. H., Garmany, J. D., and Anderson, D. L. (1973). Earthquake prediction: Variation of seismic velocities before the San Francisco earthquake, *Science 180*, 632–5. Author's correction: Read "Fernando" for "Francisco."

Wiechert, E. (1907). Ueber Erdbebenwellen. I. Theoritisches über die Ausbreitung der Erdbebenwellen, *Nachr. Ges. Wiss. Göttingen, Math. Phys. Klasse*, 415–529.

Willis, J. R. (1971). Interfacial stresses induced by arbitrary loading of dissimilar elastic half-spaces joined over a circular region, *J. Inst. Math. Applic. 7*, 179–97.

(1972). The penny-shaped crack on an interface, *Quart. J. Mech. Appl. Math. 25*, 367–85.

(1973). Self-similar problems in elastodynamics, *Phil. Trans. Roy. Soc. London A274*, 435–91.

Wong, T. F. (1982). Shear fracture energy of Westerly granite from post-failure behavior, *J. Geophys. Res. 87*, 990–1000.

Wong, T. F., and Biegel, R. (1985). Effects of pressure on the micromechanics of faulting in San Marcos gabbro, *J. Struct. Geol. 7*, 737–49.

Wyss, M. (1975). A search for precursors to the Sitka, 1972 earthquake: Sea level, magnetic field and P-residuals, *Pageoph 113*, 297–309.

Wyss, M., and Brune, J. N. (1967). The Alaska earthquake of 28 March 1964: A complex multiple rupture, *Bull. Seismol. Soc. Am. 57*, 1017–23.

(1968). Seismic moment, stress and source dynamics for earthquakes in the California–Nevada region, *J. Geophys. Res. 73*, 4681–94.

(1971). Regional variations of source properties in southern California estimated from the ratio of short- to long-period amplitudes, *Bull. Seismol. Soc. Am. 61*, 1153–67.

ADDITIONAL READING

Boldface numbers in parentheses indicate chapters in this volume to which readings apply.

Achenbach, J. D. (1974). On dynamic effects in brittle fracture, in *Mechanics Today*, ed. S. Nemat-Nasser (New York: Pergamon), Vol. 1, pp. 1–57 (**1**).

Aki, K., and Richards, P. G. (1980). *Quantitative Seismology: Theory and Methods* (San Francisco: Freeman), Chap. 12 (**3**).

Backus, G., and Gilbert, F. (1967). Numerical applications of a formalism for geophysical inverse problems, *Geophys. J. Roy. Astr. Soc. 13*, 247–76 (**3**).

(1968). The resolving power of gross earth data, *Geophys. J. Roy. Astr. Soc. 16*, 169–205 (**3**).

(1970). Uniqueness in the inversion of inaccurate gross earth data, *Phil. Trans. Roy. Soc. Lond. A266*, 123–92 (**3**).

Beran, M. J. (1968). Statistical continuum theories, in *Monographs in Statistical Physics* (New York: Wiley-Interscience), Vol. 9, Chap. 5 (**1**).

Cherepanov, G. P. (1979). *Mechanics of Brittle Fracture* (New York: McGraw-Hill) (**2**).

Dmowska, R., and Rice, J. R. (1986). Fracture theory and its seismological applications, in *Continuum Theories in Solid Earth Physics, Physics and Evolution of the Earth's Interior*, ed. R. Teisseyre (New York: Elsevier), pp. 187–255 (**2**).

Freund, L. B. (1976). The analysis of elastodynamic crack tip stress fields, in *Mechanics Today*, ed. S. Nemat-Nasser (New York: Pergamon), Vol. 3, pp. 55–91 (**2**).

(1985). The mechanics of dynamic crack growth in solids, in *Fundamentals of Deformation and Fracture*, ed. B. A. Bilby, K. J. Miller, and J. R. Willis (Cambridge University Press), pp. 163–85 (**2**).

(in press). *The Mechanics of Dynamic Fracture* (Cambridge University Press) (**2**).

Gilbert, F., and Dziewonski, A. M. (1975). An application of normal mode theory to the retrieval of structural parameters and source mechanisms from seismic spectra, *Phil. Trans. Roy. Soc. Lond. A278*, 187–269 (**3**).

Kanninen, M. F., and Popelar, C. H. (1985). *Advanced Fracture Mechanics* (New York: Oxford University Press) (**2**).

Kostrov, B. V. (1974). Crack propagation at variable velocity, *PMM (J. Appl. Math. Mech.)* *38*, 551–60 (**2**).

Lanczos, C. (1961). *Linear Differential Operators* (New York: Van Nostrand) (**3**).

Lawn, B. R., and Wilshaw, T. B. (1975). *Fracture of Brittle Solids* (Cambridge University Press) (**1**).

McCoy, J. J. (1981). Macroscopic response of continua with random microstructures, in *Mechanics Today*, ed. S. Nemat-Nasser (New York: Pergamon), Vol. 6, pp. 1–40 (**1**).

Menke, W. (1984). *Geophysical Data Analysis: Discrete Wave Theory* (New York: Academic Press) (**3**).

Orowan, E. (1960). Mechanism of seismic faulting, in *Rock Deformation*, ed. D. Griggs and J. Handin (Baltimore, MD: Geological Society of America), Chap. 12 (**1**).

INDEX